CELL BIOLOGY RESEARCH PROGRESS

NEW DEVELOPMENTS IN ASTROCYTES RESEARCH

CELL BIOLOGY RESEARCH PROGRESS

Additional books in this series can be found on Nova's website under the Series tab.

Additional e-books in this series can be found on Nova's website under the e-books tab.

CELL BIOLOGY RESEARCH PROGRESS

NEW DEVELOPMENTS IN ASTROCYTES RESEARCH

LEYLA ESTRADA
EDITOR

NOVA BIOMEDICAL

New York

Additional color graphics may be available in the e-book version of this book.

Library of Congress Cataloging-in-Publication Data

ISBN: 978-1-63485-561-7

Published by Nova Science Publishers, Inc. † New York

Contents

PREFACE

Astrocytes, the most prevalent type of cell found in the brain, are glial cells that play a critical role in the support and maintenance of homeostatic conditions in the central nervous system (CNS). In addition to regulating synaptic function and neuronal development, astrocytes contribute to CNS immune responses and regulate permeability of the blood-brain barrier. Astrocytes respond to, and become activated by, a variety of molecular events in the CNS including the appearance of abnormal molecules, the disappearance of signaling molecules that are typically present, and changes in extracellular levels of various cytokines and chemokines. Under normal physiological conditions, astrocytic activation can be advantageous, as it can help protect the CNS from pathogens or injury; however, if this activation is sustained for a prolonged period, it can lead to detrimental astrocyte reactivity. This current book provides new developments in astrocyte research. The first chapter reviews pathological roles of astrocytes in brain edema formation. Chapter Two discusses neuroimmune interactions of astrocytes with other CNS cell types. Chapter Three focuses on astrocytes in amyotrophic lateral sclerosis. The final chapter studies astrocyte's response to stress and injury.

Chapter 1 – Brain edema is a fatal pathological state that occurs after brain injuries such as stroke and head trauma, brain tumor, infection, and liver failure. In edematous brain, excess accumulation of extracellular fluid results in elevation of intracranial pressure, leading to impaired nerve function. Despite the seriousness of brain edema, only symptomatic treatments to remove edema fluid are currently available. Thus, novel anti-edema drugs to prevent the induction of brain edema are required. The pathogenesis of brain edema is classified into cytotoxic and vasogenic edema. Cytotoxic edema is characterized by swelling of astrocytes. Ion transporters and water channels in astrocytes regulate osmolality and ion composition of the brain extracellular fluid. Following brain energy depletion after ischemia or liver failure, these transporters are deregulated and a massive influx of water into astrocytes induces astrocytic swelling. Inhibition of various astrocytic transporters can reduce astrocytic swelling and cytotoxic edema in animal injury models. Vasogenic edema is defined as extracellular accumulation of fluid resulting from disruption of the blood–brain barrier (BBB) and extravasation of serum proteins, while cytotoxic edema is characterized by cell swelling caused by intracellular accumulation of fluid. Astrocytes are though to play an important role in the pathogenesis of both types of edema. The BBB largely comprises low permeability brain microvascular endothelial cells. Astrocyte processes are in close contact with brain microvascular endothelial cells, and release of astrocyte-derived factors can modulate brain microvessel permeability. Indeed, several studies have shown that excessive production of astrocytic permeability factors can cause BBB disruption and vasogenic edema after brain injury. Thus, modulation of astrocytic functions may be beneficial for the treatment of brain edema. In this chapter, the roles of key astrocytic molecules, including Na^+-K^+-Cl^- co-transporter 1, SUR1/TRPM4 channel, aquaporin-4, matrix metallo-proteinases, vascular endothelial growth factor, angiopoietins, endothelin ET_B receptors, estrogen, and progesterone, in brain edema formation are described. The pharmacological significance of astrocytes as a target of anti-brain edema drugs is also discussed.

Chapter 2 – Astrocytes, the most prevalent type of cell found in the brain, are glial cells that play a critical role in the support and maintenance of homeostatic conditions in the central nervous system (CNS). In addition to regulating synaptic function and neuronal development, astrocytes contribute to CNS immune responses and regulate permeability of the blood-brain barrier. Astrocytes respond to, and become activated by, a variety of molecular events in the CNS including the appearance of abnormal molecules, the disappearance of signaling molecules that are typically present, and changes in extracellular levels of various cytokines and chemokines. Under normal physiological conditions, astrocytic activation can be advantageous, as it can help protect the CNS from pathogens or injury; however, if this activation is sustained for a prolonged period, it can lead to detrimental astrocyte reactivity. Such sustained activation, or astrogliosis, is associated with morphological, transcriptional and secretome changes in astrocytes, including hypertrophy of the cell body, activation of intracellular signaling cascades, overexpression of the intermediate filament glial fibrillary acidic protein and increased release of pro-inflammatory cytokines. Reactive astrogliosis can lead to collateral damage to surrounding cells and potentially promote a chronic neuroinflammatory state in the brain. Due to their critical roles in the support and maintenance of CNS immune status and neurotransmission, astrocytes are considered to be one of the most complex cells in the brain. In the following chapter, the authors will describe the diverse immune interactions astrocytes have with other key cell types of the CNS, including endothelial cells, neurons, microglia and oligodendrocytes. They will also discuss the relevance and contribution of these interactions to the pathogenic mechanisms of neurodegenerative diseases.

Chapter 3 – Amyotrophic lateral sclerosis (ALS) is the third most common neurodegenerative disease, rapidly progressing and fatal. ALS is characterized by the loss of motor neurons, although neuroinflammation and glial cell activation are also part of the pathological underlying mechanisms. Indeed, phenotypically dysfunctional astrocytes with toxic properties towards motor neurons were identified in humans with either sporadic or familial ALS, as well as in transgenic rodent models, but the mechanisms involved are far from being completely clarified. Since no adequate treatment is available, intraspinal transplantation of neural stem cells to prevent motor neuron loss in ALS has been attempted to provide neuroprotection and/or replace lost cells. However, therapeutic benefits were scarce and limited to particular time frames when translated into clinical trials, which may derive from the neurotoxic properties of local astrocytes secreting unknown neurotoxins at the time of transplantation. Here, the authors summarize the current understanding on the role of astrocytes in ALS pathology, highlighting their deregulated protein signaling pathways, the associated neuroinflammatory biomarkers, as well as the gain of toxic functions along disease progression. In addition, the authors point out the contribution that soluble factors, extracellular microvesicles (ectosomes) and exosomes, as part of the astrocyte secretome, may have in disturbing normal cell-to-cell communication. Such circulating vesicles due to their content in specific proteins, mRNAs and microRNAs, have been associated to the dissemination of misfolded, mutant proteins, neurotoxins and neuroinflammatory mediators in neurodegenerative diseases, such as ALS. Finally, they address the potential benefits of astrocyte-targeted approaches for motor neuron protection in ALS, focusing on replacement-based therapies that may use astrocytes generated from induced pluripotent stem cells (iPSCs) or transdifferentiated from fibroblasts collected in patients with ALS.

Chapter 4 – Astrocytes are most abundant glial cells in the brain and their functions are widely known. Astrocytes support morphogenesis and localization of neurons in the brain through the formation of astroglial networks. Selective molecular transport between inside and outside of the central nervous system (CNS) mediated by the blood-brain barrier (BBB) is regulated by astrocytes to control the function of endothelial cells that construct the brain capillary. Astrocytes digest and remove excess neurotransmitters released from the pre-synaptic neurons for the control of neurotransmitter's concentration in the synaptic cleft. Furthermore, astrocytes produce and secrete neurotransmitter-like molecules such as glutamine and D-serine as glio-transmitters. Recently, there are a number of reports that astrocytes contribute to neuronal protection by producing antioxidants such as glutathione and metallothionein, apolipoprotein E-containing high-density lipoprotein-like particles (apoE/HDL). The authors have found that the release of fibroblast growth factor 1 (FGF-1) is enhanced from rat astrocytes exposed to the gentle oxidative stress without inducing apoptosis. The gentle oxidative stress suppresses transiently syntheses of lipids such as cholesterol, phosphatidylcholine, and sphingomyelin, and enhances lipid release from the cell surface, resulted in increasing permeability of the plasma membrane to induce release of FGF-1 in addition to the cytosolic proteins such as HSP110, HSP90, and HSP70. FGF-1 enhances the generation of apoE/HDL through the activation of at least three kinds of signaling in stress-less and healthy rat astrocytes without oxidative stress. The apoE/HDL works to protect astrocytes and neurons from oxidative stress. Astrocytes are much more sensitive to oxidative stress for the release of FGF-1 and cytosolic proteins than other cell strains. Astrocytes may have a higher performance of sensor against oxidative stress to enhance FGF-1 release than other cell strains. These findings suggest that astrocytes have a potential system using an FGF-1/apoE/HDL to protect neural cells from oxidative stress in the brain.

In: New Developments in Astrocytes Research ISBN: 978-1-63485-561-7
Editor: Leyla Estrada © 2016 Nova Science Publishers, Inc.

Chapter 1

PATHOLOGICAL ROLES OF ASTROCYTES IN BRAIN EDEMA FORMATION: A POTENTIAL TARGET OF ANTI-EDEMA DRUGS

Yutaka Koyama[*]
Laboratory of Pharmacology, Faculty of Pharmacy,
Osaka Ohtani University, Tonda-bayashi, Osaka, Japan

ABSTRACT

Brain edema is a fatal pathological state that occurs after brain injuries such as stroke and head trauma, brain tumor, infection, and liver failure. In edematous brain, excess accumulation of extracellular fluid results in elevation of intracranial pressure, leading to impaired nerve function. Despite the seriousness of brain edema, only symptomatic treatments to remove edema fluid are currently available. Thus, novel anti-edema drugs to prevent the induction of brain edema are required. The pathogenesis of brain edema is classified into cytotoxic and vasogenic edema. Cytotoxic edema is characterized by swelling of astrocytes. Ion transporters and water channels in astrocytes regulate osmolality and ion composition of the brain extracellular fluid. Following

[*] Corresponding author: Dr. Yutaka Koyama, Laboratory of Pharmacology, Faculty of Pharmacy, Osaka Ohtani University, 3-11-1 Nishikiori-Kita, Tonda-bayashi, Osaka 584-8540, Japan TEL&FAX: +81-721-24-9462 E-mail: koyamay@osaka-ohtani.ac.jp.

brain energy depletion after ischemia or liver failure, these transporters are deregulated and a massive influx of water into astrocytes induces astrocytic swelling. Inhibition of various astrocytic transporters can reduce astrocytic swelling and cytotoxic edema in animal injury models. Vasogenic edema is defined as extracellular accumulation of fluid resulting from disruption of the blood–brain barrier (BBB) and extravasation of serum proteins, while cytotoxic edema is characterized by cell swelling caused by intracellular accumulation of fluid. Astrocytes are though to play an important role in the pathogenesis of both types of edema. The BBB largely comprises low permeability brain microvascular endothelial cells. Astrocyte processes are in close contact with brain microvascular endothelial cells, and release of astrocyte-derived factors can modulate brain microvessel permeability. Indeed, several studies have shown that excessive production of astrocytic perm ebility factors can cause BBB disruption and vasogenic edema after brain injury. Thus, modulation of astrocytic functions may be beneficial for the treatment of brain edema. In this chapter, the roles of key astrocytic molecules, including Na^+-K^+-Cl^- co-transporter 1, SUR1/TRPM4 channel, aquaporin-4, matrix metallo-proteinases, vascular endothelial growth factor, angiopoietins, endothelin ET_B receptors, estrogen, and progesterone, in brain edema formation are described. The pharmacological significance of astrocytes as a target of anti-brain edema drugs is also discussed.

Keywords: brain edema, astrocyte, ion transport, blood-brain barrier, endothelial cell

INTRODUCTION

Brain edema is a pathological state often induced in the acute phase of stroke and head trauma, brain tumor, infection, and liver failure. As excessive accumulation of water in brain tissue impairs nerve function and can often lead to death, management of brain edema is clinically important. At present, intravenous injection of mannitol, diuretics, and steroids are used to decrease edematous fluid, while in some cases decompressive craniotomy is performed to protect the brain from increased intracranial pressure [1]. However, these are largely symptomatic therapies, while curative treatments to prevent the induction of brain edema have not been clinically applied. Thus, the development of novel therapeutic strategies for the improvement of brain edema is required.

Numerous experimental studies have examined the mechanisms underlying brain edema formation in individual brain disorders [2-4], with increasing evidence for a role of astrocytes [5-7]. Astrocytes regulate the composition of ions and other solutes in brain interstitial fluid by controlling intra- and extra-movement through plasma membrane ion transporters and channels. In perisynaptic areas, astrocyte processes surround spines and pre-synaptic terminals. Further, astrocytic regulation of ionic composition, especially extracellular K^+, in perisynaptic areas is responsible for the maintenance of nerve excitability. In perivascular areas, astrocytes are in close contact with brain microvessels via specialized structures termed 'end-feet' [4, 7, 8] (Figure 1). Astrocytic end-feet surround the majority of the outer surface of vascular endothelial cells with intervening basal lamina. This gliovascular complex contributes to a large part of the blood–brain barrier (BBB) [7, 8], where entry of blood components into the brain parenchyma is restricted by the low permeability of the vascular endothelial and glial layers. In the developing brain, the integrity and barrier function of microvessels is also enhanced by contact with astrocytes. Similarly, the permeability of microvessels in the mature brain is not static, but dynamically controlled under coordinated actions of various signal molecules released from astrocytes [9].

Several experimental studies have reported the presence of cerebrospinal fluid (CSF) flow in the perivascular space enclosed by vascular endothelial cells and astrocytic end-feet [10, 11], and this CSF circulation was proposed to promote supply of various factors required for function of the blood to the brain parenchyma. To maintain efficient movement of solutes between the CSF and brain interstitial fluid, astrocytic transport systems and channels are important. There is increasing evidence that impairment of solute movement between the CSF and brain interstitial fluid may be involved in various brain disorders including brain edema [10, 11]. In brain edema formation, astrocytes stimulate production of various signal molecules to control ionic movement through transporters and channels. In animal models of brain injury, manipulations of various astrocytic functions can reduce induction of brain edema, suggesting that astrocytes may be an effective target for anti-brain edema drugs. In this chapter, the roles of astrocytes in the induction of brain edema are described, and the pharmacological significance of astrocytic molecules as targets of anti-edema drugs is discussed.

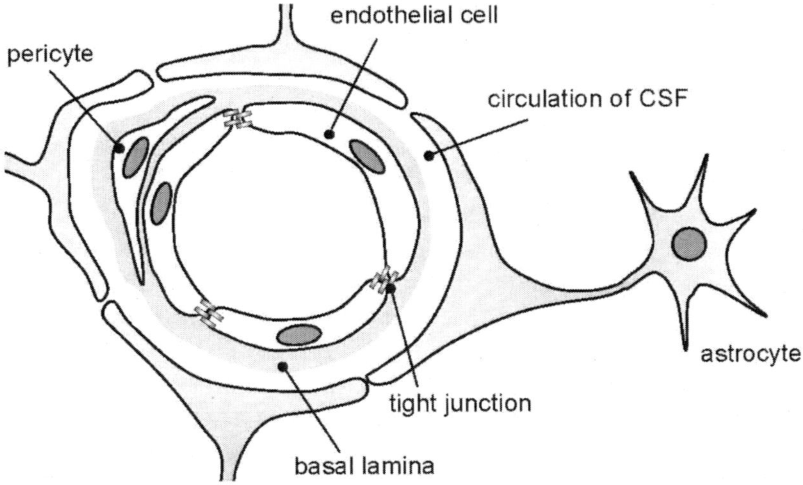

Figure 1. *Structure of gliovascular complex*. Gliovascular complex contributes to a large part of the BBB. Brain microvessels are primary composed of BMEC monolayer surrounded by basal lamina. Together with pericytes, astrocytes are in close contact with brain microvessels, where astrocytic end-feet surround the outer surface of vascular endothelial cells with intervening basal lamina. Between adjacent cells, BMECs form many TJs to be responsible for barrier functions of brain microvessels. Recent studies showed a circulation of CSF in the space enclosed by basal lamina and astrocytic end-feet, called 'glymphatic system.' On brain disorders, the CSF circulating around brain microvessels is suggested to provide Na^+ and water to brain interstitial fluid, which leads to brain edema formation.

Roles of Astrocytes in the Induction of Brain Edema

Brain edema is classically defined as cytotoxic edema and vasogenic edema [4, 10]. Although both types of brain edema can occur in brain ischemia and traumatic injury [12], they exhibit different pathogenic mechanisms. Cytotoxic edema is induced by a reduction in brain energy supply in brain ischemia, traumatic injury, infection, liver failure, and chemical intoxication, and is characterized by cellular swelling of neurons and astrocytes. By contrast, during the induction of cytotoxic edema, the BBB is intact and extravasation of blood solutes is not observed.

The laminal membranes of astrocytic end-feet enclose dendritic spines and pre-synaptic terminals. In perisynaptic areas, astrocytic ion transporters for Na^+, K^+, Cl^-, and H^+, such as Na^+/K^+-ATPase, K^+-Cl^- co-transporter-1 (KCC1), and Na^+-K^+-Cl^- co-transporter-1 (NKCC1), regulate the composition of the extracellular fluid to maintain nerve excitability. As astrocytes require consistent energy supply to maintain the activity of these ion transporters, impairment of energy metabolism in brain disorders can disable astrocytic ion transport and ionic balance, resulting in intracellular accumulation of Na^+ and Cl^- in astrocytes (Figure 2). Accumulation of Na^+ and Cl^- is followed by osmotic pressure-driven movement of water into astrocytes and leads to astrocytic swelling. In the acute phase of stroke, for example, astrocytic swelling occurs within a few minutes after the insult. Astrocytic swelling does not directly cause net accumulation of water in the brain, but the resulting decreases in interstitial volume and Na^+ concentrations allo influxes of Na^+ and water from circulating CSF into the brain, resulting in brain edema [10, 12].

Anatomical and functional studies on the gliovascular complex have proposed an important role of CSF circulation along brain microvessels. For example, Iliff et al., showed a constitutive exchange of solutes between the paravascular CSF and brain interstitial fluid, whereby CSF entering into the paravascular space of penetrating arteries (the Vichow-Robin space) is transported into brain parenchyma across barriers composed by astrocytic end-feet (glial limitation) [13, 14]. Astrocytic end-feet have several active transport systems and a relative weak intercellular adhesion compared with that of microvascular endothelial cells [7, 8]. This property allows effective supply of molecules into the brain tissue from the CSF. The brain interstitial fluid that contains metabolic waste products are removed from the brain tissue into the CSF via the paravascular space along the veins. By analogy with the peripheral lymphatic systems, this brain interstitial circulation is termed the 'glymphatic system' [11]. According to this presumed role of the glymphatic system, disturbance of exchange between the CSF and brain interstitial fluid is suggested to be involved in brain edema [10]; i.e., the massive influx of Na^+ and water into the brain parenchyma following cytotoxic edema is provided from the paravascular CSF across astrocytic end-feet. Thus, dysfunction of astrocytic transporters in end-feet largely affect the processes by which cytotoxic edema induces net accumulation of brain water.

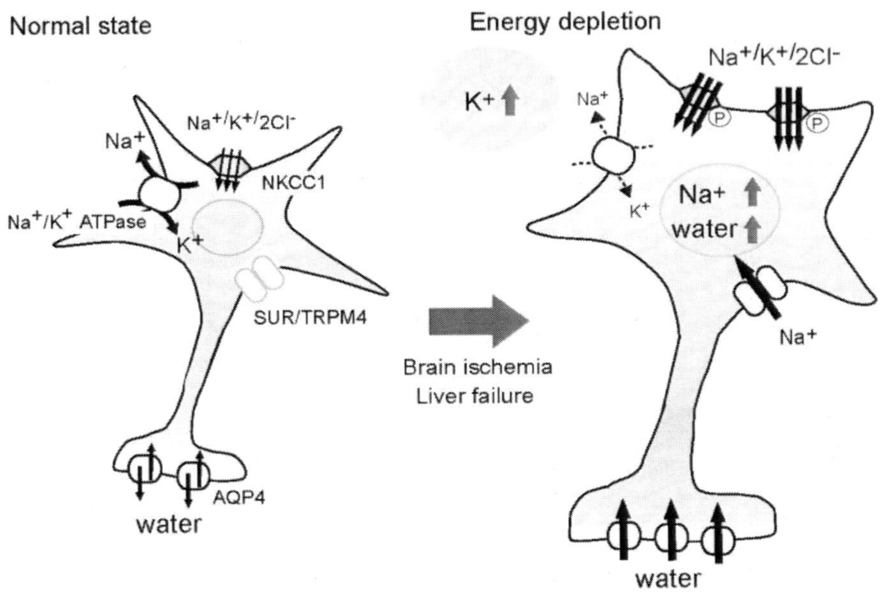

Figure 2. *Astrocytic molecules involving in induction of cytotoxic edema.* In brain ischemia and acute liver failure, ATP levels in neurons and astrocytes is decreased. Decreases in ATP levels causes reduced activity of Na^+/K^+ ATPase and increased extracellular concentrations of K^+ ($[K^+]_o$). High levels of $[K^+]_o$ stimulate NKCC1 activity, and then, intracellular accumulations of Na^+ and Cl^- allows influx of water through AQP4. These alterations cause astrocytic swelling leading to cytotoxic edema. In addition, brain disorder increases expressions of astrocytic NKCC1, SUR1/TRPM4 and AQP4 proteins, which aggravates cytotoxic edema.

Vasogenic edema is defined as that caused by extravasation of blood solutes through a disrupted BBB [4, 10]. Under physiological conditions, blood components need to pass through the brain microvessel endothelial cell (BMEC) layer when entering the brain parenchyma. Molecules required by brain cells are transferred from the blood through several pathways, including passive diffusion (lipid and non-polar molecules), transcytosis (some macromolecules), and carrier-mediated active transport in endothelial cells [7, 8]. However, macromolecules and polar molecules without endothelial transporters are generally blocked by the endothelial cell layer. Thus, these solutes enter through extracellular gaps between adjacent BMECs. BMECs show firm intercellular contact supported by a large numbers of adhesion junctions (AJs). In addition, plasma membrane proteins associated with tight junctions (TJs, occludin, and claudins) are highly expressed in BMECs [6, 8].

The characteristic abundance of AJs and TJs is largely responsible for preventing entry of blood solutes through endothelial gaps. However, this intercellular barrier function of BMECs is not static, but dynamically altered by interactions with astrocytes [7, 9] (Figure 3).

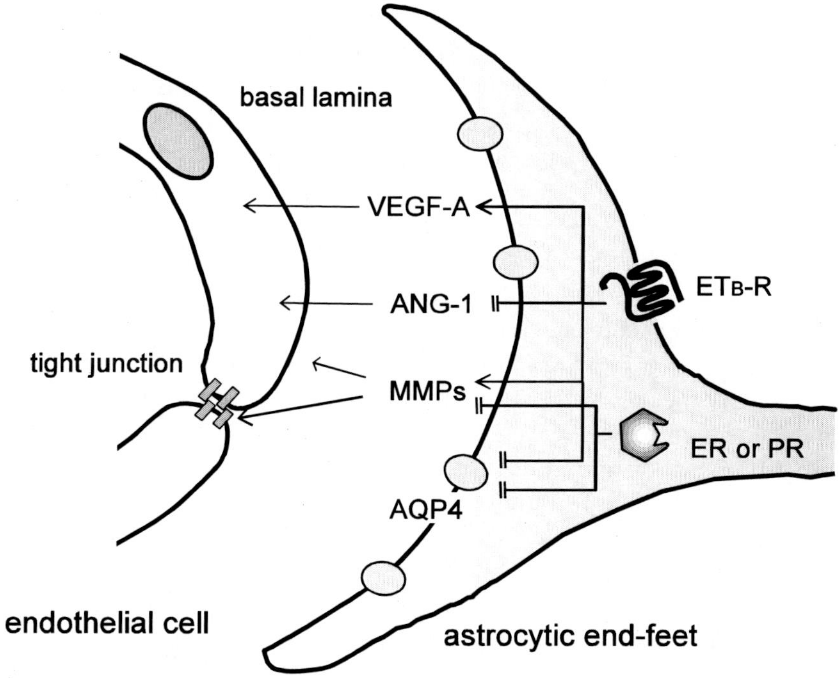

Figure 3. *Astrocytic molecules involving in induction of vasogenic edema.* At gliovascular complex, permeability of BBB is dynamically regulated under a balance of astrocyte-derived factors. However, brain disorders often shifts this balance to excessive reduction of the barrier function, and causes vasogenic edema. On brain disorders, astrocytes became reactive. Reactive astrocytes increase productions of MMPs, which reduces integrity of BMEC layers by degeneration of basal lamina and TJ proteins. In addition, production of astrocytic VEGF-A is also increased, while ANG-1 is decreased. These changes in astrocyte-derived factor productions reduced barrier function of BMECs. This mechanism of vasogenic edema formation indicates a key role of reactive astrocytes. In fact, some studies showed that disruption of BBB and vasogenic edema formation were attenuated by estradiol, progesterone and ET_B antagonists, which prevent induction of reactive astrocytes. ET_B-R: endothelin ET_B receptor, ER: estrogen receptor, PR: progesterone receptor.

During angiogenesis in the repair processes of injured nerve tissue, astrocytes produce soluble factors that enhance the barrier function of BMECs. Such astrocytic factors include glia-derived neurotrophic factor, fibroblast growth factor-2, angiopoietin-1 (ANG-1), and Src-suppressed C kinase substrate. By contrast, astrocytes also produce factors that enhance endothelial permeability, including matrix metalloproteinasees (MMPs), inflammatory cytokines, chemokines, and vascular endothelial growth factor (VEGF). Thus, the permeability of BMECs is physiologically maintained under a balance of these vascular permeability-regulating factors produced by astrocytes. The production of permeability-regulating factors in astrocytes is also affected by signaling factors derived from BMECs, suggesting that astrocyte-BMCE interactions are likely important in the regulation of BBB function [7]. In many brain disorders, astrocytes convert their phenotype to reactive astrocytes [15, 16], characterized by cellular hypertrophy and increased expression of GFAP. With this phenotypic conversion, astrocytes also alter their functions to induce pathophysiological responses of injured nerve tissues [9, 15, 16]. In damaged areas, the balanced regulation of BMEC permeability by astrocytes is often impaired, where excessive production of permeability-enhancing factors cause BBB disruption and extravasation of blood solutes into the brain. Thus, the functional alteration of astrocytes in brain disorders plays a pivotal role in the induction of vasogenic edema.

Possible Targets of Anti-Edema Drugs in Astrocytes

With advances in our understanding of the mechanisms of cytotoxic and vasogenic edema, the importance of astrocytic regulation of the ionic composition of the brain interstitial fluid and BBB permeability is now widely recognized. Numerous studies have focused on the identification and characterization of key molecules involved in astrocytic regulatory functions. A number of molecules have been identified to play a role in brain edema formation, and modulation of their functions can effectively prevent brain edema and nerve injury in various brain disorders. Such astrocytic molecules that are expected to be novel targets of anti-brain edema drugs are detailed below.

Na^+-K^+-Cl^- co-transporter-1 (NKCC1)

The NKCC transporter is a membrane protein that controls unidirectional co-transport of Na^+, K^+, and Cl^- across the plasma membrane [17]. NKCC1 has two isoforms, NKCC1 and NKCC2. NKCC2 is exclusively expressed in the Henle loop of renal tubules and controls reabsorption of Na^+, K^+, and Cl^- from primary urine. By contrast, NKCC1 is widely present in various types of cells. In astrocytes, the major physiological function of NKCC1 involves control of intracellular influx of Na^+, K^+, and Cl^- required for regulation of extracellular K^+ concentrations [18]. NKCC1-mediated ion transport is a secondary active transport driven by the transmembrane Na^+ gradient formed by the Na^+/K^+ ATPase. A number of studies have reported that astrocytic NKCC1 activity is stimulated in various brain disorders and that increased ion transport can cause astrocytic swelling [19, 20]. In brain energy depletion induced by ischemia and acute liver failure, a reduction of Na^+/K^+ ATPase-mediated Na^+ and K^+ antiporters resulted in increased extracellular K^+ concentration ($[K^+]_o$). The increased $[K^+]_o$ stimulates astrocytic NKCC1 activity and allows intracellular accumulation of Na^+ and Cl^-, which causes astrocytic swelling and cytotoxic edema [4, 21]. In addition, brain acidosis in ischemic tissue stimulates NKCC1 activity, likely via activation of Na^+/H^+ exchangers. Chen et al., reported that NKCC1-null mice showed reduced brain edema formation and reduced infarction volume after MCAO-induced ischemic injury [22]. Administration of bumetanide, a loop diuretic that inhibits NKCCs, effectively attenuated brain edema in animal models of stroke and liver failure [19, 23, 24]. These studies suggest that inhibition of NKCC1 may be a promising strategy to prevent cytotoxic edema.

The key roles of NKCC1-mediated ion transport in astrocytic swelling were also reported in cultured astrocytes [20, 25]. NKCC1 protein expression is up-regulated in various brain disorders, while enhancement of NKCC1 activity may aggravate brain edema. As expression of NKCC1 in cultured astrocytes was stimulated by hypoxia, proinflammatory cytokines (tumor necrosis factor-α, interleukin-1β), and NH_4^+ [26, 27], these factors are suggested to be involved in the induction of cytotoxic edema. In addition, recent studies reported the regulation of astrocytic NKCC1 activity by protein phosphorylation [28, 29]. Further studies on the regulatory mechanisms of astrocytic NKCC1 may provide a novel strategy to prevent cytotoxic edema by suppressing NKCC1-mediated ion transport.

SUR1/TRPM4 Cation Channel

Sulfonylurea receptor 1 (SUR1) is a membrane protein associated with a pore-forming subunit of Kir6.x found in heterooctametric K_{ATP} channels [30]. SUR1 serves as a regulatory subunit of K_{ATP} channels, while this function is inhibited by sulfonylurea anti-diabetic meditcations. SUR1 also regulates the opening of Ca^{2+}-activated ATP-sensitive non-selective cation (NCca-ATP) channels [31, 32], a transient receptor potential cation channel superfamily M member 4 (TRPM4) [33]. Depolarizing currents through TRPM4 channels were observed in reactive astrocytes freshly prepared from the injured adult brain, while the cation influx was accompanied by astrocytic swelling [32, 34]. Expression of astrocytic SUR1 and TRPM4 were also up-regulated in stroke, subarachnoid hemorrhage (SAH), head trauma, and acute liver failure [35-38]. Glibenclamide, an inhibitor of SUR1, was reported to reduce ATP depletion-induced opening of TRPM4 and cell swelling in cultured astrocytes [31, 39], indicating a functional coupling of SUR1 and TRPM4. These findings suggest pathophysiological roles of SUR1/TRPM4 channels in the formation of cytotoxic brain edema. Simord et al., also reported a reduction in brain edema and infarction volume with glibenclamide treatment in a mouse brain ischemia injury model [32]. Further, attenuation of brain edema and neural damage by inhibition of SUR1 was observed in animal models of head trauma, SAH, and acute liver failure [39-41]. The preclinical efficacy of treatment with SUR1 inhibitors is currently been investigated in a clinical trial on stroke patients [42].

Aquaporin-4 (AQP4)

AQP4, a family of water channel proteins consisting of 13 members, regulates water movement across the plasma membrane driven by osmotic and hydrostatic pressure gradients. AQP4 is the principle AQP in the brain, and its location is largely restricted to end-feet of perivascular astrocytes, which constitute the glial limitation between the paravascular CSF and brain parenchyma extracellular fluid [43, 44]. As water movement across the plasma membrane at end-feet can directly affect astrocytic swelling and resulting brain edema [4, 10, 12], the specific localization of AQP4 suggests a potential role in brain edema formation [45, 46]. Observations on AQP4 null-mice showed a reduction in brain ischemia- or water intoxication-induced cytotoxic brain edema [47], indicating that water entry through astrocytic AQP4 causes cytotoxic brain edema. Further, AQP4 deletion resulted in reduced brain edema in traumatic brain injury and acute liver failure models [48, 49].

AQP4 was also proposed to play a role in vasogenic edema formation. Papadopulos et al., showed that continuous intracerebral fluid infusion caused higher intracranial pressure and water content in AQP4 null mice compared with wild-type mice [50]. Moreover, AQP4-null mice showed higher intracranial pressure and water content in vasogenic brain edema after cerebral cold-injury and tumor implantation, suggesting that AQP4 acts as an excretion pathway for edematous fluid from the brain [51]. Although experimental data suggest that AQP4 is an interesting target of anti-brain edema drugs, agents that specifically inhibit AQP4 have not been found. The creation of specific AQP4 inhibitors will provide an important step towards the development of effective treatments for brain edema.

MMPs

MMPs are a Zn^{2+}-endopeptidase family responsible for the degradation of extracellular matrix proteins such as collagen, laminin, and fibronectin. The proteolytic activity of MMPs is required for tissue remodeling processes in many pathological conditions, including cell proliferation, migration, and angiogenesis [52, 53]. During repair of impaired nervous tissues, angiogenesis is initiated by MMP-mediated degradation of the basal lamina and TJ proteins around brain microvessels [54]. In astrocytes, MMP2 and MMP14 (membrane type-MMP1) are constitutively expressed at end-feet [55], where the pro-form of MMP2 is activated by cleavage of its pro-peptide domain by MMP14 [56]. MMP3 and MMP9 are also produced by astrocytes, although their expression levels in the normal brain are low. Production of astrocytic MMP3 and MMP9 is up-regulated in response to brain injury [57-60], and pro-MMP9 is activated by MMP3 and free radicals. Increased activation of astrocytic MMP2 and MMP9 around the gliovascular complex causes degeneration of basal lamina and endothelial TJ proteins, thus reducing the integrity of the BMEC layer.

While reduction of endothelial integrity is essential for initiation of vascular remodeling, excessive activation of MMP2 and MMP9 can impair the barrier function of BMECs and allow extravasation of blood solutes leading to induction of vasogenic edema [53, 54]. Yang and Rosenberg described a detailed time-course of MMP activation and BBB disruption in an animal model of stroke [61]. In the early phase (within a few hours) after reperfusion, proteolytic activity of MMP2 was stimulated by the action of hypoxia-activated MMP14 and furin, while increased MMP2 activity was correlated with degeneration of TJ proteins (claudin-5 and occludin) and transient enhancement of vascular permeability. In the delayed phase (24–72 h) after reperfusion, MMP3 and MMP9 activity were predominantly increased, which induced further degradation of the basal lamina and TJ proteins [62-64]. The coordinated increase in MMP2 and MMP9 activity results in BBB disruption, extravasation of blood solutes, infiltration of inflammatory cells, and vasogenic edema. MMP2 and MMP9 are also involved in vasogenic edema formation induced by head trauma, SAH, and brain tumors. Thus, inhibition of MMP activity is considered a therapeutic strategy to prevent vasogenic edema.

Based on this therapeutic strategy, several MMP inhibitors, including GM6001, BB-1101, KB-R7785, SB-3CT, and MMI270, were reported to attenuate brain edema and neural damage in animal models of brain injury, including ischemia, head trauma, and SAH [65-69].

VEGF-A

As described above, astrocytes release various signaling molecules that regulate the permeability of BMECs [7, 9, 15]. For example, VEGF-A, a prototype of VEGF isopeptides, stimulates proliferation and migration of vascular endothelial cells [70, 71], and can promote angiogenesis in adult nervous tissue repair. During angiogenesis, the integrity of pre-existing blood vessels needs to be reduced to initiate proliferation and migration of vascular endothelial cells. VEGF-A can also reduce endothelial integrity in pathological conditions [70, 71]. Studies on VEGF-induced vascular remodeling show that activation of endothelial VEGF receptors can weaken barrier function of TJs by inducing cytoskeletal actin reorganization, phosphorylation and down-regulation of TJ proteins [72-74]. As reduction of BMEC integrity is followed by enhancement of vascular permeability, VEGF-A is a potent edema-inducing factor [71]. Increased VEGF levels in the serum and brain were reported in SAH and brain trauma patients [75, 76]. Experimentally, stimulation of astrocytic VEGF-A production was also observed in various animal models of brain injury [77-79]. Further, in brain areas damaged by ischemia, head trauma, and brain tumors, VEGF-A production correlated with BBB disruption [80, 81], suggesting that excessive VEGF-A activity caused vasogenic edema formation. In support, administration of VEGF inhibitors (VEGF-A neutralizing antibody, soluble Flt-1, and receptor tyrosine kinase inhibitors) attenuated BBB disruption and brain edema induced by brain ischemia and cold injury [82-85]. In the brain, BMCEs and subpopulations of neurons produce VEGF-A. However, Argaw et al., showed a predominance of astrocyte-derived VEGF-A production in brain edema formation, whereby astrocyte-specific deletion of VEGF-A reduced BBB disruption in a mouse multiple sclerosis model [86].

By contrast, inhibition of VEGF-A during the later phase of stroke may actually delay repair of damaged nerve tissue by preventing angiogenesis. Indeed, delayed administration of VEGF-A promoted angiogenesis and protected neurons from ischemic and traumatic injury [87-89]. VEGF receptors are also present in neurons and neural progenitors. Neural VEGF receptors can produce anti-apoptotic and neurogenic effects of VEGF [90-92], indicating potential neuroprotective actions of increased VEGF-A production in various brain disorders. Thus, although inhibition of VEGF-A is a promising strategy to prevent vasogenic edema, further studies are required to determine the most effective time window for treatment with VEGF inhibitors.

Angiopoietin-1 (ANG-1)

Angiopoietins (ANGs) are another growth factor family that can modulate vascular permeability [93, 94]. Among the ANG proteins, ANG-1 and ANG-2 are well characterized for their actions on vascular endothelial cells. Both ANG-1 and ANG-2 are specific ligands for a tyrosine kinase-linked receptor, Tie-2, although they produce different effects on vascular endothelial cell function. ANG-1, which is constitutively expressed in the brain, induces autophosphorylation of Tie-2 to modulate the physiological functions of BMECs [95]. Histological observations show that ANG1 is produced by astrocytes and pericytes surrounding the brain microvessels [96, 97]. Activation of endothelial Tie-2 causes tightening of cell adhesion between BMECs by increasing expression of cell adhesion proteins [98]. By contrast, expression levels of ANG-2 are normally low in the brain, but increased in various brain disorders [99, 100]. ANG-2 acts as an endogenous antagonist of ANG-1 on Tie-2, thus inducing apoptosis of BMECs [100]. In this manner, the integrity and permeability of BMECs are maintained under balance of ANG-1 and ANG-2.

Alterations in ANG-1 and ANG-2 expression have been examined in brain ischemia and cold injury models. For example, expression of ANG-1 was transiently decreased, while ANG-2 was increased, during BBB disruption [80, 99]. Administration of ANG-1 also reduced BBB disruption and brain edema induced by brain ischemia and traumatic injury [96, 101, 102]. These studies suggest that targeting ANG-1 signaling may be useful for reducing brain edema. A few signaling molecules such as Src-suppressed C kinase substrate, glucocorticoids, and endothelin-1 (ET-1) were reported to regulate astrocytic ANG-1 production [103-105]. Although the altered ANG-1 production is discussed in relation to BMEC permeability, the regulatory mechanisms for astrocytic ANG production remain largely unknown.

ET_B Receptor

Phenotypic conversion of astrocytes to a reactive state can occur in many brain disorders [15, 16]. As conversion to reactive astrocytes is accompanied by increased production of edema-inducing factors, such as MMPs and VEGF-A [7, 9], then inhibition of reactive astrocytes is expected to reduce BBB disruption and brain edema. ETs can induce conversion to reactive astrocytes through ET_B receptors. For example, administration of an ET_B agonist into the rat brain increased the number of reactive astrocytes [106, 107], while BQ788, an ET_B antagonist, reduced induction of reactive astrocytes after brain injury [108, 109]. In support, inhibitory effects of ET_B antagonists on brain edema formation have been reported. Administration of ET_B antagonists (BQ788 and IRL2500) into the ventricles improved BBB disruption and brain edema in rat cerebral ischemia and cold brain injury models [85, 110, 111]. Expression of MMP9 and VEGF-A were also decreased with the reduction of brain edema with the use of an ET_B antagonist [85]. In the rat brain and in cultured astrocytes, ET-1 also stimulated MMP9 and VEGF-A production through ET_B receptors [105, 112, 113]. These studies suggest that blockade of astrocytic ET_B receptors in brain disorders may be an effective strategy to prevent vasogenic brain edema by decreasing edema-inducing factor productions.

Activation of ET_B receptors can also increase MMP2 and MMP3 production in astrocytes [113, 114]. Moreover, ETs can down-regulate astrocytic AQP4 proteins through ET_B receptors [115, 116], which may aggravate brain edema by closing edematous fluid removal pathways. Inhibition of these actions of ET-1 on astrocytes may be involved in the anti-edema effects of ET_B antagonists. As described, MMPs and VEGF-A are astrocytic factors that induce vasogenic edema [61, 71], and inhibition of these factors in brain disorders have anti-edema effects. However, as brain edema involves the synergistic actions of many edema-inducing factors [7, 9], inhibition of a single factor is likely insufficient to effectively prevent brain edema. Nevertheless, blockade of astrocytic ET_B receptors can extensively attenuate the actions of multiple edema-inducing factors, which may be an advantage in the use of ET_B antagonists as anti-brain edema drugs.

Estradiol and Progesterone

The neuroprotective actions of gonadal hormone have been suggested by clinical data showing sex differences in the incident rate, severity of brain failure, and response rate to therapeutics of acute brain injury and neurodegenerative diseases [117, 118]. Many studies in animal brain injury models have shown neuroprotective actions of estradiol and progesterone [119, 120]. The proposed protective mechanisms include promotion of neuronal survival and regeneration, reduction of neuroinflammation, BBB stabilization, and improvement of brain edema. Although both of the nuclear estrogen receptor subtypes (ERα and ERβ) are present in astrocytes, astrocytic ERα expression was up-regulated by brain injury [121] and could affect astrocytic function [122]. Further, administration of 17-β-estradiol decreased the number of reactive astrocytes after brain injury [123-125]. As the negative regulation of astrocytic phenotypic conversion is accompanied by a reduction in the production of edema-inducing factors, this may underlie the anti-edema action of 17-β-estradiol. Using astrocyte-specific deletion of ERs, Spence et al., showed that the ERα receptor in astrocytes, but not in neurons, mediated the neuroprotective action of 17-β-estradiol observed in mouse experimental autoimmune encephalomyelitis [126]. Treatment with 17-β-estradiol also decreased astrocytic AQP4 expression, coincident with a reduction in cellular swelling [127-129]. Inhibitory effects of estradiol on the production of astrocytic inflammatory cytokines and MMP9 were also reported [130].

Inhibitory actions of progesterone on brain edema and BBB disruption were observed in animal models of brain ischemia, traumatic injury, and SAH [119, 120]. As for estradiol, progesterone negatively regulates the induction of reactive astrocytes [123, 131] and decreases expressions of astrocytic AQP4 [129, 132]. Ishrat et al., reported that brain ischemia-induced MMP9, MMP2, and VEGF production was decreased by administration of progesterone [133]. Based on these preclinical findings, clinical trials of progesterone for traumatic brain injury have also been performed. In phase 2 randomized trials, administration of progesterone after brain insult caused a reduction in mortality and improved functional outcome [134, 135]. However, the following phase 3 multicenter trials failed to indicate significant differences between progesterone and placebo groups [136]. Recent studies on neuroprotective actions of progesterone also suggest a role for membrane-bound progesterone receptors [137, 138], although their functions remain unclear. Further studies on the pathophysiological roles of each progesterone receptor subtype may resolve the contrasting results between preclinical and clinical data.

CONCLUSION

Although brain edema is a fatal state occurring in various brain disorders, effective treatments to prevent edema formation have not been established. There is now an increase in the development of therapeutic strategies that target astrocytes for the treatment of neurodegenerative and mental diseases [9, 139, 140]. Similarly, as the regulatory roles of astrocytes in the maintenance of the BBB and glymphatic system become more widely understood, astrocytes are becoming a promising target of anti-edema drugs

[4, 5, 9]. Many astrocytic molecules have been found to play a role in aggravating brain edema, some of which already have specific inhibitors or activators that are used clinically, including sulfonylurea anti-diabetics, loop diuretics, ET antagonists, and gonadal hormones. In the future, these drugs may be examined in clinical trials or aid in the development of novel anti-edema drugs.

ACKNOWLEDGMENTS

This work was supported by a Grant-in-Aid for Scientific Research (C) from the Japan Society for the Promotion of Science (15K07981). The author has no conflicts of interest related to this study.

REFFERENCES

[1] Raslan, A. & Bhardwaj, A. (2007). Medical management of cerebral edema. *Neurosurg Focus.*, *22*, E12.

[2] Bosoi, C. R. & Rose, C. F. (2013). Brain edema in acute liver failure and chronic liver disease: similarities and differences. *Neurochem Int.*, *62*, 446-457.

[3] Unterberg, A. W., Stover, J., Kress, B. & Kiening, K. L. (2004). Edema and brain trauma. *Neuroscience.*, *129*, 1021-1029.

[4] Walcott, B. P., Kahle, K. T. & Simard, J. M. (2012). Novel treatment targets for cerebral edema. *Neurotherapeutics.*, *9*, 65-72.

[5] Shields, J., Kimbler, D. E., Radwan, W., Yanasak, N., Sukumari-Ramesh, S. & Dhandapani, K. M. (2011). Therapeutic targeting of astrocytes after traumatic brain injury. *Transl Stroke Res.*, *2*, 633-642.

[6] Quaegebeur, A., Lange, C. & Carmeliet, P. (2011). The neurovascular link in health and disease: molecular mechanisms and therapeutic implications. *Neuron.*, *71*, 406-424.

[7] Alvarez, J. I., Katayama, T. & Prat, A. (2013). Glial influence on the blood brain barrier. *Glia*, *61*, 1939-1958.

[8] Wolburg, H., Noel, l. S., Mack, A., Wolburg-Buchholz, K. & Fallier-Becker, P. (2009). Brain endothelial cells and the glio-vascular complex. *Cell Tissue Res.*, *335*, 75-96.

[9] Buffo, A., Rolando, C. & Ceruti, S. (2010). Astrocytes in the damaged brain: molecular and cellular insights into their reactive response and healing potential. *Biochem Pharmacol.*, *79*, 77-89.

[10] Thrane, A. S., Rangroo Thrane, V. & Nedergaard, M. (2014). Drowning stars: reassessing the role of astrocytes in brain edema. *Trends Neurosci.*, *37*, 620-628.

[11] Jessen, N. A., Munk, A. S., Lundgaard, I. & Nedergaard, M. (2015). The Glymphatic System: A Beginner's Guide. *Neurochem Res.*, *40*, 2583-2599.

[12] Simard, J. M., Kent, T. A., Chen, M., Tarasov, K. V. & Gerzanich, V. (2007). Brain oedema in focal ischaemia: molecular pathophysiology and theoretical implications. *Lancet Neurol.*, *6*, 258-268.

[13] Iliff, J. J., Wang, M., Liao, Y., Plogg, B. A., Peng, W., Gundersen, G. A., Benveniste, H., Vates, G. E., Deane, R., Goldman, S. A., Nagelhus, E. A. & Nedergaard, M. (2012). A paravascular pathway facilitates CSF flow through the brain parenchyma and the clearance of interstitial solutes, including amyloid β. *Sci Transl Med.*, *4*, 147ra111.

[14] Iliff, J. J., Lee, H., Yu, M., Feng, T., Logan, J., Nedergaard, M. & Benveniste, H. (2013). Brain-wide pathway for waste clearance captured by contrast-enhanced MRI. *J Clin Invest.*, *123*, 1299-309.

[15] Sofroniew, M. V. (2009). Molecular dissection of reactive astrogliosis and glial scar formation. *Trends Neurosci.*, *2*, 638-647.

[16] Koyama, Y. (2014). Signaling molecules regulating phenotypic conversions of astrocytes and glial scar formation in damaged nerve tissues. *Neurochem Int.*, *78*, 35-42.

[17] Haas, M. (1994). The Na-K-Cl cotransporters. *Am J Physiol.*, *267*, C869-885.

[18] Jayakumar, A. R. & Norenberg, M, D. (2010). The Na-K-Cl Co-transporter in astrocyte swelling. *Metab Brain Dis.*, *25*, 31-38.

[19] Yan, Y., Dempsey, R. J. & Sun D. (2001). Na+-K^+-Cl^- cotransporter in rat focal cerebral ischemia. *J Cereb Blood Flow Metab,*. *21*, 711-721.

[20] Su, G., Kintner, D. B., Flagella, M., Shull, G. E. & Sun, D. (2002). Astrocytes from Na^+-K^+-Cl- cotransporter-null mice exhibit absence of swelling and decrease in EAA release. *Am J Physiol Cell Physiol.*, *282*, C11470-1160.

[21] Reinert, M., Khaldi, A., Zauner, A., Doppenberg, E., Choi, S. & Bullock, R. (2000), High extracellular potassium and its correlates after severe head injury: relationship to high intracranial pressure. *Neurosurg Focus.*, *8*, e10.

[22] Chen, H. & Sun, D. (2005). The role of Na-K-Cl co-transporter in cerebral ischemia *Neurol Res.*, *27*, 280-286.

[23] Chen, H., Luo, J., Kintner, D. B., Shull, G. E. & Sun, D. (2005). Na^{+}-dependent chloride transporter (NKCC1)-null mice exhibit less gray and white matter damage after focal cerebral ischemia. *J Cereb Blood Flow Metab.*, *25*, 54-66.

[24] Jayakumar, A. R., Valdes, V. & Norenberg, M. D. (2011). The Na-K-Cl cotransporter in the brain edema of acute liver failure. *J Hepatol.*, *54*, 272-278.

[25] Jayakumar, A. R., Panickar, K. S., Curtis, K. M., Tong, X. Y., Moriyama, M. & Norenberg, M. D. (2011). Na-K-Cl cotransporter-1 in the mechanism of cell swelling in cultured astrocytes after fluid percussion injury. *J Neurochem.*, *117*, 437-448.

[26] Huang, L. Q., Zhu, G. F., Deng, Y. Y., Jiang, W. Q., Fang, M., Chen, C. B., Cao, W., Wen, M. Y., Han, Y. L. & Zeng, H. K. (2014). Hypertonic saline alleviates cerebral edema by inhibiting microglia-derived TNF-α and IL-1β-induced Na-K-Cl Cotransporter up-regulation. *J Neuroinflammation.*, *11*, 102.

[27] Sinke, A. P., Jayakumar, A. R., Panickar, K. S., Moriyama, M., Reddy, P. V. & Norenberg, M. D. (2008). NFκB in the mechanism of ammonia-induced astrocyte swelling in culture. *J Neurochem.*, *106*, 2302-2311.

[28] Jayakumar, A. R., Panickar, K. S., Murthy, C. R. & Norenberg, M. D. (2006). Oxidative stress and mitogen-activated protein kinase phosphorylation mediate ammonia-induced cell swelling and glutamate uptake inhibition in cultured astrocytes. *J Neurosci.*, *26*, 4774-4784.

[29] Jayakumar, A. R., Liu, M., Moriyama, M., Ramakrishnan, R., Forbush, B. 3rd, Reddy, P. V. & Norenberg, M. D. (2008). Na-K-Cl Cotransporter-1 in the mechanism of ammonia-induced astrocyte swelling. *J Biol Chem.*, *283*, 33874-33882.

[30] Seino, S. 1. ATP-sensitive potassium channels: a model of heteromultimeric potassium channel/receptor assemblies. *Annu Rev Physiol.*, 1999, *61*, 337-62.

[31] Chen, M., Dong, Y. & Simard, J. M. (2003). Functional coupling between sulfonylurea receptor type 1 and a nonselective cation channel in reactive astrocytes from adult rat brain. *J Neurosci.*, *23*, 8568-8577.

[32] Simard, J. M., Chen, M., Tarasov, K. V., Bhatta, S., Ivanova, S., Melnitchenkom L., Tsymbalyuk, N., Wes, G. A. & Gerzanich, V. (2006). Newly expressed SUR1-regulated $NC_{Ca\text{-}ATP}$ channel mediates cerebral edema after ischemic stroke. *Nat Med.*, *12*, 433-440.

[33] Woo, S. K., Kwon, M. S., Ivanov, A., Gerzanich, V. & Simard, J. M. (2013). The sulfonylurea receptor 1 (Sur1)-transient receptor potential melastatin 4 (Trpm4) channel. *J Biol Chem.*, *288*, 3655-3667.

[34] Chen, M. & Simard, J. M. (2001). Cell swelling and a nonselective cation channel regulated by internal Ca^{2+} and ATP in native reactive astrocytes from adult rat brain. *J Neurosci.*, *21*, 6512-6521.

[35] Simard, J. M., Woo, S. K., Schwartzbauer, G. T. & Gerzanich, V. (2012). Sulfonylurea receptor 1 in central nervous system injury: a focused review. *J Cereb Blood Flow Metab.*, *32*, 1699-1717.

[36] Martínez-Valverde, T., Vidal-Jorge, M., Martínez-Saez, E., Castro, L., Arikan, F., Cordero, E., Rădoi, A., Poca, M. A., Simard, J. M. & Sahuquillo, J. (2015). Sulfonylurea receptor 1 in humans with post-traumatic brain contusions. *J Neurotrauma.*, *32*, 1478-1487.

[37] Jayakumar, A. R., Valdes, V., Tong, X. Y., Shamaladevi, N., Gonzalez, W. & Norenberg, M. D. (2014). Sulfonylurea receptor 1 contributes to the astrocyte swelling and brain edema in acute liver failure. *Transl Stroke Res.*, *5*, 28-37.

[38] Mehta, R. I., Tosun, C., Ivanova, S., Tsymbalyuk, N., Famakin, B. M., Kwon, M. S., Castellani, R. J., Gerzanich, V. & Simard, J. M. (2015). Sur1-Trpm4 Cation Channel Expression in Human Cerebral Infarcts. *J Neuropathol Exp Neurol.*, *74*, 835-849.

[39] Jayakumar, A. R1., Valdes, V., Tong, X. Y., Shamaladevi, N., Gonzalez, W. & Norenberg, M. D. (2014). Sulfonylurea receptor 1 contributes to the astrocyte swelling and brain edema in acute liver failure. *Transl Stroke Res.*, Feb, 5(1), 28-37.

[40] Zweckberger, K., Hackenberg, K., Jung, C. S., Hertle, D. N., Kiening, K. L., Unterberg, A. W. & Sakowitz, O. W. (2014). Glibenclamide reduces secondary brain damage after experimental traumatic brain injury. *Neuroscience.*, *272*, 199-206.

[41] Tosun, C., Kurland, D. B., Mehta, R., Castellani, R. J., deJong, J. L., Kwon, M. S., Woo, S. K., Gerzanich, V. & Simard, J. M. (2013). Inhibition of the Sur1-Trpm4 channel reduces neuroinflammation and cognitive impairment in subarachnoid hemorrhage. *Stroke.*, *44*, 3522-3528.

[42] Sheth, K. N., Kimberly, W. T., Elm, J. J., Kent, T. A., Yoo, A. J., Thomalla, G., Campbell, B., Donnan, G. A., Davis, S. M., Albers, G. W., Jacobson, S., del Zoppo, G., Simard, J. M., Stern, B. J. & Mandava, P. (2014). Exploratory analysis of glyburide as a novel therapy for preventing brain swelling. *Neurocrit Care.*, *21*, 43-51.

[43] Amiry-Moghaddam, M. & Ottersen, O. P. (2003). The molecular basis of water transport in the brain. *Nat Rev Neurosci.*, *4*, 991-1001.

[44] Tait, M. J., Saadoun, S., Bell, B. A. & Papadopoulos, M. C. (2008). Water movements in the brain: role of aquaporins. *Trends Neurosci.*, *31*, 37-43.

[45] Nielsen, S., Nagelhus, E. A., Amiry-Moghaddam, M., Bourque, C., Agre, P. & Ottersen, O. P. (1997). Specialized membrane domains for water transport in glial cells: high-resolution immunogold cytochemistry of aquaporin-4 in rat brain. *J Neurosci.*, *17*, 171-180.

[46] Rash, J. E., Yasumura, T., Hudson, C. S., Agre, P. & Nielsen, S. (1998). Direct immunogold labeling of aquaporin-4 in square arrays of astrocyte and ependymocyte plasma membranes in rat brain and spinal cord. *Proc Natl Acad Sci USA.*, *95*, 11981-11986.

[47] Manley, G. T., Fujimura, M., Ma, T., Noshita, N., Filiz, F., Bollen, A. W., Chan, P. & Verkman, A. S. (2000). Aquaporin-4 deletion in mice reduces brain edema after acute water intoxication and ischemic stroke. *Nat Med.*, *6*, 159-163.

[48] Yao, X., Uchida, K., Papadopoulos, M. C., Zador, Z., Manley, G. T. & Verkman, A. S. (2015). Mildly reduced brain swelling and improved neurological outcome in aquaporin-4 knockout mice following controlled cortical impact brain Injury. *J. Neurotrauma.*, *32*, 1458-1464.

[49] Rama Rao, K. V., Verkman, A. S., Curtis, K. M. & Norenberg, M. D. (2014). Aquaporin-4 deletion in mice reduces encephalopathy and brain edema in experimental acute liver failure. *Neurobiol Dis.*, *63*, 222-228.

[50] Papadopoulos, M. C., Manley, G. T., Krishna, S. & Verkman, A. S. (2004). Aquaporin-4 facilitates reabsorption of excess fluid in vasogenic brain edema. *FASEB J.*, *18*, 1291-1293.

[51] Saadoun, S. & Papadopoulos, M. C. (2010). Aquaporin-4 in brain and spinal cord oedema. *Neuroscience.*, *168***,** 1036-1046.

[52] Mott, J. D. & Werb, Z. (2004). Regulation of matrix biology by matrix metalloproteinases. *Curr Opin Cell Biol.*, *16*, 558–564.

[53] Dzwonek, J., Rylski, M. & Kaczmarek, L. (2004). Matrix metalloproteinases and their endogenous inhibitors in neuronal physiology of the adult brain. *FEBS Lett*, *567*, 129-135.

[54] Candelario-Jalil, E., Yang, Y. & Rosenberg, G. A. (2009). Diverse roles of matrix metalloproteinases and tissue inhibitors of metalloproteinases in neuroinflammation and cerebral ischemia. *Neuroscience.*, *158*, 983-994.

[55] Yong, V. W., Power, C., Forsyth, P. & Edwards, D. R. (2001). Metalloproteinases in biology and pathology of the nervous system. *Nat Rev Neurosci.*, *2*, 502-511.

[56] English, J. L., Kassiri, Z., Koskivirta, I., Atkinson, S. J., Di Grappa, M., Soloway, P. D., Nagase, H., Vuorio, E., Murphy, G. & Khokha, R. (2006). Individual Timp deficiencies differentially impact pro-MMP-2 activation. *J Biol Chem.*, *281*, 10337-10346.

[57] Yang, Y., Jalal, F. Y., Thompson, J. F., Walker, E. J., Candelario-Jalil, E., Li, L., Reichard, R. R., Ben, C., Sang, Q. X., Cunningham, L. A. & Rosenberg, G. A. (2011). Tissue inhibitor of metalloproteinases-3 mediates the death of immature oligodendrocytes via TNF-α/TACE in focal cerebral ischemia in mice. *J Neuroinflammation.*, *8*, 108.

[58] Gertten, C., Holmin, S., Mathiesen, T. & Nordqvist, A. C. (2003). Increases in matrix metalloproteinase-9 and tissue inhibitor of matrix metalloproteinase-1 mRNA after cerebral contusion and depolarization. *J Neurosci Res.*, *73*, 803-810.

[59] Adair, J. C., Charlie, J., Dencoff, J. E., Kaye, J. A., Quinn, J. F., Camicioli, R. M., Stetler-Stevenson, W. G. & Rosenberg, G. A. (2004). Measurement of gelatinase B (MMP-9) in the cerebrospinal fluid of patients with vascular dementia and Alzheimer disease. *Stroke.*, *35*, 159-162.

[60] Solé, S., Petegnief, V., Gorina, R., Chamorro, A. & Planas. A. M. (2004). Activation of matrix metalloproteinase-3 and agrin cleavage in cerebral ischemia/reperfusion. *J Neuropathol Exp Neurol.*, *63*, 38-349.

[61] Yang, Y. & Rosenberg, G. A. (2015). Matrix metalloproteinases as therapeutic targets for stroke. *Brain Res.*, *1623*, 30-38.

[62] Bauer, A. T., Bürgers, H. F., Rabie, T. & Marti, H. H. (2010). Matrix metalloproteinase-9 mediates hypoxia-induced vascular leakage in the brain via tight junction rearrangement. *J Cereb Blood Flow Metab. 30*, 837-848.

[63] Liu, J., Jin, X., Liu, K. J. & Liu, W. (2012). Matrix metalloproteinase-2-mediated occludin degradation and caveolin-1-mediated claudin-5 redistribution contribute to blood-brain barrier damage in early ischemic stroke stage. *J Neurosci.*, *32*, 3044-3057.

[64] Yang, Y., Estrada, E. Y., Thompson, J. F., Liu, W. & Rosenberg, G. A. (2007). Matrix metalloproteinase-mediated disruption of tight junction proteins in cerebral vessels is reversed by synthetic matrix metalloproteinase inhibitor in focal ischemia in rat. *J Cereb Blood Flow Metab.*, *27*, 697-709.

[65] Wang, J. & Tsirka, S. E. (2005). Neuroprotection by inhibition of matrix metalloproteinases in a mouse model of intracerebral haemorrhage. *Brain.*, *128*, (Pt 7), 1622-1633.

[66] Jiang, X., Namura, S. & Nagata, I. (2001). Matrix metalloproteinase inhibitor KB-R7785 attenuates brain damage resulting from permanent focal cerebral ischemia in mice. *Neurosci Lett.*, *305*, 41-44.

[67] Jin, X., Liu, J., Yang, Y., Liu, K. J., Yang, Y. & Liu, W. (2012). Spatiotemporal evolution of blood brain barrier damage and tissue infarction within the first 3h after ischemia onset. *Neurobiol Dis.*, *48*, 309-316.

[68] Jia, F., Yin, Y. H., Gao, G. Y., Wang, Y., Cen, L. & Jiang, J. Y. (2014). MMP-9 inhibitor SB-3CT attenuates behavioral impairments and hippocampal loss after traumatic brain injury in rat. *J Neurotrauma.*, *31*, 1225-1234.

[69] Kawai, N., Kawanishi, M., Okada, M., Matsumoto, Y. & Nagao, S. (2003). Treatment of cold injury-induced brain edema with a nonspecific matrix metalloproteinase inhibitor MMI270 in rats. *J Neurotrauma.*, *20*, 649-657.

[70] Roskoski, R. Jr. (2008). VEGF receptor protein-tyrosine kinases: structure and regulation. *Biochem Biophys Res Commun.*, *375*, 287-291.

[71] Bates, D. O. (2010). Vascular endothelial growth factors and vascular permeability. *Cardiovasc Res.*, *87*, 262-271.

[72] Argaw, A. T., Gurfein, B. T., Zhang, Y., Zameer, A. & John, G. R. (2009). VEGF-mediated disruption of endothelial CLN-5 promotes blood-brain barrier breakdown. *Proc. Natl. Acad. Sci. USA*, *6*, 977–982.

[73] Nico, B., Mangieri, D., Crivellato, E., Longo, V., De Giorgis, M., Capobianco, C., Corsi, P., Benagiano, V., Roncali, L. & Ribatti, D. (2007). HIF activation and VEGF overexpression are coupled with ZO-1 up-phosphorylation in the brain of dystrophic mdx mouse. *Brain Pathol.*, *7*, 399-406.

[74] Wang, W., Dentler, W. L. & Borchardt, R. T. (2001). VEGF increases BMEC monolayer permeability by affecting occludin expression and tight junction assembly. *Am J Physiol Heart Circ Physiol.*, *280*, H434-440.

[75] Gong, D., Zhang, S., Liu, L., Dong, J., Guo, X., Hao, M., Tu, Y., Diao, Y. & Zhang, J. (2011). Dynamic changes of vascular endothelial growth factor and angiopoietin-1 in association with circulating endothelial progenitor cells after severe traumatic brain injury. *J Trauma.*, *70*, 1480-1484.

[76] Mellergård, P., Sjögren, F. & Hillman, J. (2010). Release of VEGF and FGF in the extracellular space following severe subarachnoidal haemorrhage or traumatic head injury in humans. *Br J Neurosurg., 24*, 261-267.

[77] Hayashi, T., Abe, K., Suzuki, H. & Itoyama, Y. (1997). Rapid induction of vascular endothelial growth factor gene expression after transient middle cerebral artery occlusion in rats. *Stroke, 28*, 2039-2044.

[78] Sköld, M. K., von Gertten, C., Sandberg-Nordqvist, A. C., Mathiesen, T. & Holmin, S. (2005). VEGF and VEGF receptor expression after experimental brain contusion in rat. *J Neurotrauma. 22*, 353-367.

[79] Papavassiliou, E., Gogate, N., Proescholdt, M., Heiss, J. D., Walbridge, S., Edwards, N. A., Oldfield, E. H. & Merrill, M. J. (1997). Vascular endothelial growth factor (vascular permeability factor) expression in injured rat brain. *J Neurosci Res., 49*, 451-460.

[80] Zhang, Z. G., Zhang, L., Tsang, W., Soltanian-Zadeh, H., Morris, D., Zhang, R., Goussev, A., Powers, C., Yeich, T. & Chopp, M. (2002). Correlation of VEGF and angiopoietin expression with disruption of blood-brain barrier and angiogenesis after focal cerebral ischemia. *J Cereb Blood Flow Metab., 22*, 379-392.

[81] Zhang, Z. G., Zhang, L., Jiang, Q., Zhang, R., Davies, K., Powers, C., Bruggen, N. & Chopp, M. (2000). VEGF enhances angiogenesis and promotes blood-brain barrier leakage in the ischemic brain. *J Clin Invest., 106*, 829-838.

[82] Van Bruggen, N., Thibodeaux, H., Palmer, J. T., Lee, W. P., Fu, L., Cairns, B., Tumas, D., Gerlai, R., Williams, S. P. & van Lookeren, C. M. et al., (1999). VEGF antagonism reduces edema formation and tissue damage after ischemia/reperfusion injury in the mouse brain. *J. Clin. Investig., 104*, 1613–1620.

[83] Kimura, R., Nakase, H., Tamaki, R. & Sakaki, T. (2005). Vascular endothelial growth factor antagonist reduces brain edema formation and venous infarction. *Stroke., 36*, 1259-1263.

[84] Kumai, Y., Ooboshi, H., Ibayashi, S., Ishikawa, E., Sugimori, H., Kamouchi, M., Kitazono, T., Egashira, K. & Iida, M. (2007). Postischemic gene transfer of soluble Flt-1 protects against brain ischemia with marked attenuation of blood-brain barrier permeability. *J. Cereb. Blood Flow Metab., 6*, 1152–1160.

[85] Michinaga, S., Seno, N., Fuka, M., Yamamoto, Y., Minami, S., Kimura, A., Hatanaka, S., Nagase, M., Matsuyama, E., Yamanaka, D. & Koyama Y. (2015). Improvement of cold injury-induced mouse brain edema

by endothelin ET_B antagonists is accompanied by decreases in matrixmetalloproteinase 9 and vascular endothelial growth factor-A. *Eur J Neurosci.*, *42*, 2356-2370.

[86] Argaw, A. T., Asp, L., Zhang, J., Navrazhina, K., Pham, T., Mariani, J. N., Mahase, S., Dutta, D. J., Seto, J., Kramer, E. G., Ferrara, N., Sofroniew, M. V. & John, G. R. (2012). Astrocyte-derived VEGF-A drives blood-brain barrier disruption in CNS inflammatory disease. *J Clin Invest.*, *122*, 2454-2468.

[87] Sun, Y., Jin, K., Xie, L., Childs, J., Mao, X. O., Logvinova, A. & Greenberg, D. A. (2003). VEGF-induced neuroprotection, neurogenesis, and angiogenesis after focal cerebral ischemia. *J Clin Invest.*, *111*, 1843-1851.

[88] Wang, Y. Q., Cui, H. R., Yang, S. Z., Sun, H. P., Qiu, M. H., Feng, X. Y. & Sun, F. Y. (2009). VEGF enhance cortical newborn neurons and their neurite development in adult rat brain after cerebral ischemia. *Neurochem Int.*, *55*, 629-636.

[89] Figley, S. A., Liu, Y., Karadimas, S. K., Satkunendrarajah, K., Fettes, P., Spratt, S. K., Lee, G., Ando, D., Surosky, R., Giedlin, M. & Fehlings, M. G. (2014). Delayed administration of a bio-engineered zinc-finger VEGF-A gene therapy is neuroprotective and attenuates allodynia following traumatic spinal cord injury. *PLoS One.*, *9*, e96137.

[90] Kirby, E. D., Kuwahara, A. A., Messer, R. L. & Wyss-Coray, T. (2015). Adult hippocampal neural stem and progenitor cells regulate the neurogenic niche by secreting VEGF. *Proc Natl Acad Sci U S A.*, *112*, 4128-4133.

[91] Jin, K., Mao, X. O. & Greenberg, D. A. (2006). Vascular endothelial growth factor stimulates neurite outgrowth from cerebral cortical neurons via Rho kinase signaling. *J Neurobiol.*, *66*, 236-242.

[92] Svensson, B., Peters, M., König, H. G., Poppe, M., Levkau, B., Rothermundt, M., Arolt, V., Kögel, D. & Prehn, J. H. (2002). Vascular endothelial growth factor protects cultured rat hippocampal neurons against hypoxic injury via an antiexcitotoxic, caspase-independent mechanism. *J Cereb Blood Flow Metab.*, *22*, 1170-1175.

[93] Brindle, N. P., Saharinen, P. & Alitalo, K. (2006). Signaling and functions of angiopoietin-1 in vascular protection. *Circ Res.*, *98*, 1014-1023.

[94] Shim, W. S., Ho, I. A. & Wong, P. E. (2007). Angiopoietin: a TIE(d) balance in tumor angiogenesis. *Mol Cancer Res.*, *5*, 655-665.

[95] Chittiboina, P., Ganta, V., Monceaux, C. P., Scott, L. K., Nanda, A. & Alexander, J. S. (2013). Angiopoietins as promising biomarkers and potential therapeutic targets in brain injury. *Pathophysiology.*, *20*, 15-21.

[96] Kawamura, K., Takahashi, T., Kanazawa, M., Igarashi, H., Nakada, T., Nishizawa, M. & Shimohata, T. (2014). Effects of angiopoietin-1 on hemorrhagic transformation and cerebral edema after tissue plasminogen activator treatment for ischemic stroke in rats. *PLoS One.*, *9*, e98639.

[97] Zacharek, A., Chen, J., Cui, X., Li, A., Li, Y., Roberts, C., Feng, Y., Gao, Q. & Chopp, M. (2007). Angiopoietin1/Tie2 and VEGF/Flk1 induced by MSC treatment amplifies angiogenesis and vascular stabilization after stroke. *J Cereb Blood Flow Metab.*, *27*, 1684-1691.

[98] Yu, H.1., Wang, P., An, P. & Xue, Y. Recombinant human angiopoietin-1 ameliorates the expressions of ZO-1, occludin, VE-cadherin, and PKCα signaling after focal cerebral ischemia/reperfusion in rats. *J Mol Neurosci.*, 2012 Jan, *46*(1), 236-47.

[99] Nourhaghighi, N., Teichert-Kuliszewska, K., Davis, J., Stewart, D. J. & Nag, S. (2003). Altered expression of angiopoietins during blood-brain barrier breakdown and angiogenesis. *Lab Invest.*, *83*, 1211-1222.

[100] Nag, S., Papneja, T., Venugopalan, R. & Stewart, D. J. (2005). Increased angiopoietin2 expression is associated with endothelial apoptosis and blood-brain barrier breakdown. *Lab Invest.*, *85*, 1189-1198.

[101] Thurston, G., Suri, C., Smith, K., McClain, J., Sato, T. N., Yancopoulos, G. D. & McDonald, D. M. (1999). Leakage-resistant blood vessels in mice transgenically overexpressing angiopoietin-1. *Science.*, 1 *286*, 2511-2514.

[102] Valable, S., Montaner, J., Bellail, A., Berezowski, V., Brillault, J., Cecchelli, R., Divoux, D., Mackenzie, E. T., Bernaudin, M., Roussel, S. & Petit, E. (2005). VEGF-induced BBB permeability is associated with an MMP-9 activity increase in cerebral ischemia: both effects decreased by Ang-1. *J Cereb Blood Flow Metab.*, *25*, 1491-1504.

[103] Lee, S. W., Kim, W. J., Choi, Y. K., Song, H. S., Son, M. J., Gelman, I. H., Kim, Y. J. & Kim, K. W. (2003). SSeCKS regulates angiogenesis and tight junction formation in blood-brain barrier. *Nat Med.*, *9*, 900-906.

[104] Kim, H., Lee, J. M., Park, J. S., Jo, S. A., Kim, Y. O., Kim, C. W. & Jo, I. (2008). Dexamethasone coordinately regulates angiopoietin-1 and VEGF: a mechanism of glucocorticoid-induced stabilization of blood-brain barrier. *Biochem Biophys Res Commun.*, *372*, 243-248.

[105] Koyama, Y., Maebara, Y., Hayashi, M., Nagae, R., Tokuyama, S. & Michinaga, S. (2012). Endothelins reciprocally regulate VEGF-A and angiopoietin-1 production in cultured rat astrocytes: Implications on astrocytic proliferation. *Glia*, *60*, 1954–1963.

[106] Ishikawa, N., Takemura, M., Koyama, Y., Shigenaga, Y., Okada, T. & Baba, A. (1997). Endothelins promote the activation of astrocytes in rat neostriatum through ETB receptors. *Eur. J. Neurosci.*, *9*, 895–901.

[107] Koyama, Y., Tsujikawa, K., Matsuda, T. & Baba, A. (2003). Intracerebroventricular administration of an endothelin ET_B receptor agonist increases expressions of GDNF and BDNF in rat brain. *Eur. J. Neurosci.*, *18*, 887–894.

[108] Koyama, Y., Takemura, M., Fujiki, K., Ishikawa, N., Shigenaga, Y. & Baba, A. (1999). BQ788, an endothelin ET_B receptor antagonist, attenuates stab wound injury-induced reactive astrocytes in rat brain. *Glia.*, *26*, 268-271.

[109] Gadea, A., Schinelli, S. & Gallo, V. (2008). Endothelin-1 regulates astrocyte proliferation and reactive gliosis via a JNK/c-Jun signaling pathway. *J Neurosci.*, *28*, 2394-2408.

[110] Michinaga, S., Nagase, M., Matsuyama, E., Yamanaka, D., Seno, N., Fuka, M., Yamamoto, Y. & Koyama, Y. (2014). Amelioration of cold injury-induced cortical brain edema formation by selective endothelin ET_B receptor antagonists in mice. *PLoS ONE*, *9*, e102009.

[111] Moldes, O., Sobrino, T., Blanco, M., Agulla, J., Barral, D., Ramos-Cabrer, P. & Castillo, J. (2012). Neuroprotection afforded by antagonists of endothelin-1 receptors in experimental stroke. *Neuropharmacology*, *63*, 1279-1285.

[112] Koyama, Y., Nagae, R., Tokuyama, S. & Tanaka, K. (2011). I.c.v. administration of an endothelin ET_B receptor agonist stimulates vascular endothelial growth factor-A production and activates vascular endothelial growth factor receptors in rat brain. *Neuroscience*, *192*, 689–698.

[113] Koyama, Y. & Tanaka, K. (2010). Intracerebroventricular administration of an endothelin ET_B-receptor agonist increases expression of matrix metalloproteinase-2 and -9 in rat brain. *J. Pharmacol. Sci.*, *114*, 433–443.

[114] Koyama, Y. & Tanaka, K. (2008). Endothelins stimulate the production of stromelysin-1 in cultured rat astrocytes. *Biochem Biophys Res Commun.*, *371*, 6596-63.

[115] Koyama, Y. & Tanaka, K. (2010). Decreases in rat brain aquaporin-4 expression following intracerebroventricular administration of an endothelin ET_B receptor agonist. *Neurosci Lett.*, *469*, 343-347.

[116] Tanaka, K. & Koyama, Y. (2011). Endothelins decrease the expression of aquaporins and plasma membrane water permeability in cultured rat astrocytes. *J Neurosci Res.*, *89*, 320-328.

[117] Stein, D. G. (2001). Brain damage, sex hormones and recovery: a new role for progesterone and estrogen? *Trends Neurosci.*, *24*, 386-391.

[118] Vagnerova, K., Koerner, I. P. & Hurn, P. D. (2008) Gender and the injured brain. *Anesth Analg*. *107*, 201-214.

[119] Stein, D. G. (2008). Progesterone exerts neuroprotective effects after brain injury. *Brain Res Rev.*, *57*, 386-397.

[120] Johann, S. & Beyer, C. (2011). Neuroprotection by gonadal steroid hormones in acute brain damage requires cooperation with astroglia and microglia. *J Steroid Biochem Mol Biol.*, *137*, 71-81.

[121] García-Ovejero, D., Veiga, S., García-Segura, L. M. & Doncarlos, L. L. (2002). Glial expression of estrogen and androgen receptors after rat brain injury. *J Comp Neurol.*, *450*, 256-271.

[122] Carbonaro, V., Caraci, F., Giuffrida, M. L., Merlo, S., Canonico, P. L., Drago, F., Copani, A. & Sortino, M. A. (2009). Enhanced expression of ERα in astrocytes modifies the response of cortical neurons to β-amyloid toxicity. *Neurobiol Dis.*, *33*, 415-421.

[123] Garcia-Estrada, J., Del Rio, J. A., Luquin, S., Soriano, E. & Garcia-Segura, L. M. (1993). Gonadal hormones down-regulate reactive gliosis and astrocyte proliferation after a penetrating brain injury. *Brain Res.*, *628*, 271-278.

[124] Barreto, G., Santos-Galindo, M., Diz-Chaves, Y., Pernía, O., Carrero, P., Azcoitia, I. & Garcia-Segura, L. M. (2009). Selective estrogen receptor modulators decrease reactive astrogliosis in the injured brain: effects of aging and prolonged depletion of ovarian hormones. *Endocrinology.*, *150*, 5010-5015.

[125] López Rodríguez, A. B., Mateos, Vicente, B., Romero-Zerbo, S. Y., Rodriguez-Rodriguez, N., Bellini, M. J., Rodriguez de Fonseca, F., Bermudez-Silva, F. J., Azcoitia, I., Garcia-Segura, L. M. & Viveros, M. P. (2011). Estradiol decreases cortical reactive astrogliosis after brain injury by a mechanism involving cannabinoid receptors. *Cereb Cortex.*, *21*, 2046-2055.

[126] Spence, R. D., Hamby, M. E., Umeda, E., Itoh, N., Du, S., Wisdom, A. J., Cao, Y., Bondar, G., Lam, J., Ao, Y., Sandoval, F., Suriany, S.,

Sofroniew, M. V. & Voskuhl, R. R. (2011). Neuroprotection mediated through estrogen receptor-alpha in astrocytes. *Proc Natl Acad Sci U S A.*, *108*, 8867-8872.

[127] Tomás-Camardiel, M., Venero, J. L., Herrera, A. J., De Pablos, R. M., Pintor-Toro, J. A., Machado, A. & Cano, J. (2005). Blood-brain barrier disruption highly induces aquaporin-4 mRNA and protein in perivascular and parenchymal astrocytes: protective effect by estradiol treatment in ovariectomized animals. *J Neurosci Res.*, *80*, 235-246.

[128] Rutkowsky, J. M., Wallace, B. K., Wise, P. M. & O'Donnell, M. E. (2011). Effects of estradiol on ischemic factor-induced astrocyte swelling and AQP4 protein abundance. *Am J Physiol Cell Physiol.*, *301*, C204-212.

[129] Habib, P., Dang, J., Slowik, A., Victor, M. & Beyer, C. (2014). Hypoxia-induced gene expression of aquaporin-4, cyclooxygenase-2 and hypoxia-inducible factor 1α in rat cortical astroglia is inhibited by 17β-estradiol and progesterone. *Neuroendocrinology.*, *99*, 156-167.

[130] Lewis, D. K., Johnson, A. B., Stohlgren, S., Harms, A. & Sohrabji, F. (2008). Effects of estrogen receptor agonists on regulation of the inflammatory response in astrocytes from young adult and middle-aged female rats. *J Neuroimmunol.*, *195*, 47-59.

[131] Djebaili M. 1., Guo Q, Pettus E. H., Hoffman S. W. & Stein, D. G. (2005). The neurosteroids progesterone and allopregnanolone reduce cell death, gliosis, and functional deficits after traumatic brain injury in rats. *J Neurotrauma.*, Jan, 22(1), 106-18.

[132] He, L., Zhang, X., Wei, X. & Li, Y. (2014). Progesterone attenuates aquaporin-4 expression in an astrocyte model of ischemia/reperfusion. *Neurochem Res.*, *39*, 2251-2261.

[133] Ishrat, T., Sayeed, I., Atif, F., Hua, F. & Stein, D. G. (2010). Progesterone and allopregnanolone attenuate blood-brain barrier dysfunction following permanent focal ischemia by regulating the expression of matrix metalloproteinases. *Exp Neurol.*, *226*, 183-190.

[134] Wright, D. W. 1., Kellermann, A. L., Hertzberg, V. S., Clark, P. L., Frankel, M, Goldstein, F. C., Salomone, J. P., Dent, L. L., Harris O. A., Ander, D. S., Lowery, D. W., Patel, M. M., Denson, D. D., Gordon, A. B., Wald, M. M., Gupta, S., Hoffman, S. W. & Stein, D. G. (2007). ProTECT: a randomized clinical trial of progesterone for acute traumatic brain injury. *Ann Emerg Med.*, Apr, *49*(4), 391-402.

[135] Xiao, G. 1., Wei, J., Yan, W., Wang, W. & Lu, Z. (2008). Improved outcomes from the administration of progesterone for patients with acute

severe traumatic brain injury: a randomized controlled trial. *Crit Care.*, *12*(2), R61.

[136] Skolnick, B. E., Maas, A. I., Narayan, R. K., van der Hoop, R. G., MacAllister, T., Ward, J. D., Nelson, N. R., Stocchetti, N. & SYNAPSE Trial Investigators. (2014). A clinical trial of progesterone for severe traumatic brain injury. *N Engl J Med.*, *371*, 2467-2476.

[137] Meffre, D. 1., Labombarda, F., Delespierre, B., Chastre, A., De Nicola, A. F., Stein, D. G., Schumacher, M. & Guennoun, R. (2013). Distribution of membrane progesterone receptor alpha in the male mouse and rat brain and its regulation after traumatic brain injury. *Neuroscience.*, Feb *12*, 231, 111-24.

[138] Ishrat, T., Sayeed, I., Atif, F., Hua, F. & Stein, D. G. (2012). Progesterone is neuroprotective against ischemic brain injury through its effects on the phosphoinositide 3-kinase/protein kinase B signaling pathway. *Neuroscience.*, *210*, 442-450.

[139] Colangelo, A. M., Alberghina, L. & Papa, M. (2014). Astrogliosis as a therapeutic target for neurodegenerative diseases. *Neurosci Lett.*, *565*, 59-64.

[140] Koyama, Y. (2015). Functional alterations of astrocytes in mental disorders: pharmacological significance as a drug target. *Front Cell Neurosci.*, *9*, 261.

BIOGRAPHICAL SKETCH

Name: Yutaka Koyama

Affiliation:

Laboratory of Pharmacology, Faculty of Pharmacy,
Osaka Ohtani University, Nishiki-rori 3-11-1, Tondabayashi, 584-8540, Japan

Education:

1994 Ph. D (Pharmaceutical Sciences). Graduate School of Pharmaceutical Sciences, Osaka University

1985 Bachelor of Pharmacy, Faculty of Pharmacy, Osaka University

Address: koyamay@osaka-ohtani.ac.jp

Research and Professional Experience:

2006 – present: Professor

Laboratory of Pharmacology, Faculty of Pharmacy, Osaka Ohtani University

2006 – present: Invited Professor
Graduate School of Pharmaceutical Sciences, Osaka University
2001 – 2006: Associate Professor
Laboratory of Medicinal Pharmacology, Graduate School of Pharmaceutical Sciences, Osaka University
1997 – 1998: Invited Researcher
Department of Pathology, the College of Physicians and Surgeons, Columbia University
1989 – 2001: Assistant Professor
Laboratory of Neuropharmacology, Graduate School of Pharmaceutical Sciences, Osaka University

Professional Appointments:

2012 – present: The Journal of Pharmacological Sciences, Advisory Board
2010 – present: The Neurochemistry International, Editorial Advisory Board

Honors:

1990 the Academic Award of the Japanese Pharmacological Society

Publications Last Three Years:

1. Koyama Y. (2015) Functional alterations of astrocytes in mental disorders: pharmacological significance as a drug target (Review). *Front Cell Neurosci.* **9:** 261.
2. Michinaga S, Seno N, Fuka M, Yamamoto Y, Minami S, Kimura A, Hatanaka S, Nagase M, Matsuyama E, Yamanaka D, Koyama Y. (2015) Improvement of cold injury-induced mouse brain edema by endothelin ET_B antagonists is accompanied by decreases in matrixmetalloproteinase 9 and vascular endothelial growth factor-A. *Eur J Neurosci.* **42**: 2356-2370
3. Michinaga S. and Koyama Y. (2015) Pathogenesis of brain edema and investigation into anti-edema drugs (Review). *Int J Mol Sci.* **16**: 9949-9975.
4. Michinaga S. and Koyama Y. (2015) Extravasation of endogenous serum albumin into the mouse brain. *Bio-protocol.* **5:** 1463.
5. Nakamoto K, Nishinaka T, Sato N, Aizawa F, Yamashita T, Mankura M, Koyama Y, Kasuya F, Tokuyama S, (2015) The activation of supraspinal GPR40 signaling regulates the descending pain control system. *Br J Pharmacol.*, **172**: 1250-1262.
6. Koyama Y. (2014) Signaling molecules regulating phenotypic conversions of astrocytes and glial scar formation in damaged nerve tissues (Review). *Neurochem Int.* **78**: 35-42.

7. Koyama Y, Hayashi M, Nagae R, Tokuyama S, Konishi T. (2014) Endothelin-1 increases the expression of VEGF-R1/Flt-1 receptors in rat cultured astrocytes through ET_B receptors. *J Neurochem.* **130**: 759-769.
8. Michinaga S, Nagase M, Matsuyama E, Yamanaka D, Seno N, Fuka M, Yamamoto Y, Koyama Y. (2014) Amelioration of cold injury-induced cortical brain edema formation by selective endothelin ET_B receptor antagonists in mice. *PLoS One.* **9(7)**, e102009.
9. Koyama Y, Kotani M, Sawamura T, Kuribayashi M, Konishi R, Michinaga S.(2013) Different actions of endothelin-1 on chemokine production in rat cultured astrocytes: reduction of CX3CL1/fractalkine and an increase in CCL2/MCP-1 and CXCL1/CINC-1. *J Neuroinflammation.* **10**: 51.
10. Koyama Y. (2013) Endothelin systems in the brain: involvement in pathophysiological responses of damaged nerve tissues (Review) *Biomol. Concepts*, **4**: 335-347.
11. Michinaga S, Ishida A, Takeuchi R, Koyama Y. (2013) Endothelin-1 stimulates cyclin D1 expression in rat cultured astrocytes via activation of Sp1. *Neurochem Int.*, **63**: 25-34.
12. Nakamoto K, Nishinaka T, Sato N, Mankura M, Koyama Y, Kasuya F, Tokuyama S. (2013) Hypothalamic GPR40 signaling activated by free long chain fatty acids suppresses CFA-induced inflammatory chronic pain. *PLoS One.*, **8(12)**: e81563.

In: New Developments in Astrocytes Research ISBN: 978-1-63485-561-7
Editor: Leyla Estrada

Chapter 2

NEUROIMMUNE INTERACTIONS OF ASTROCYTES WITH OTHER CENTRAL NERVOUS SYSTEM CELL TYPES

C. B. Pointer, W. T. Slattery, L. J. Spielman, J. McKenzie, J. R. Lowry, C. Lee and A. Klegeris [*]
Department of Biology,
University of British Columbia Okanagan Campus, Kelowna,
British Columbia, Canada

ABSTRACT

Astrocytes, the most prevalent type of cell found in the brain, are glial cells that play a critical role in the support and maintenance of homeostatic conditions in the central nervous system (CNS). In addition to regulating synaptic function and neuronal development, astrocytes contribute to CNS immune responses and regulate permeability of the blood-brain barrier. Astrocytes respond to, and become activated by, a variety of molecular events in the CNS including the appearance of abnormal molecules, the disappearance of signaling molecules that are typically present, and changes in extracellular levels of various cytokines and chemokines. Under normal physiological conditions, astrocytic

[*] Corresponding author: Andis Klegeris, DPhil, Department of Biology, University of British Columbia Okanagan Campus, Kelowna, British Columbia, V1V 1V7, CANADA, andis.klegeris@ubc.ca, (P) 250-807-9557, (F) 250-807-8830.

activation can be advantageous, as it can help protect the CNS from pathogens or injury; however, if this activation is sustained for a prolonged period, it can lead to detrimental astrocyte reactivity. Such sustained activation, or astrogliosis, is associated with morphological, transcriptional and secretome changes in astrocytes, including hypertrophy of the cell body, activation of intracellular signaling cascades, overexpression of the intermediate filament glial fibrillary acidic protein and increased release of pro-inflammatory cytokines. Reactive astrogliosis can lead to collateral damage to surrounding cells and potentially promote a chronic neuroinflammatory state in the brain.

Due to their critical roles in the support and maintenance of CNS immune status and neurotransmission, astrocytes are considered to be one of the most complex cells in the brain. In the following chapter, we will describe the diverse immune interactions astrocytes have with other key cell types of the CNS, including endothelial cells, neurons, microglia and oligodendrocytes. We will also discuss the relevance and contribution of these interactions to the pathogenic mechanisms of neurodegenerative diseases.

Abbreviations

AD	Alzheimer's disease
ALS	amyotrophic lateral sclerosis
AMPAR	α-amino-3-hydroxy-5-methyl-4-isoxazolepropionic acid receptor
AP	action potential
AQP4	aquaporin-4
Aβ	amyloid β
BBB	blood-brain barrier
BDNF	brain-derived neurotrophic factor
APP	amyloid precursor protein
Ca^{2+}	calcium
CBF	cerebral blood flow
CCL5	chemokine (C-C motif) ligand 5
CD54	cluster of differentiation 54
CNS	central nervous system
CNTF	ciliary neurotrophic factor
EC	endothelial cell
ET-1	endothelin-1
GABA	gamma-aminobutyric acid

GDNF	glial-derived neurotrophic factor
GFAP	glial fibrillary acidic protein
GLT-1	glutamate transporter-1
GLUT1	glucose transporter-1
HD	Huntington's disease
ICAM-1	intercellular adhesion molecule 1
IFN	interferon
IGF-1	insulin-like growth factor-1
IKK	IκB kinase
IL	interleukin
K^+	potassium
LAT	L-system amino acid transporter
LPS	lipopolysaccharide
LTD	long-term depression
LTP	long-term potentiation
MAO-B	monoamine oxidase-B
MCP-1	monocyte chemotactic protein-1
MEGF10	multiple EGF-l like domains 10
MERTK	MER proto-oncogene tyrosine kinase
mHTT	mutant huntingtin
MPP^+	1-methyl-4-phenylpyridinium
MPTP	1-methyl-4-phenyl-1, 2, 3, 6-tetrahydropyridine
MS	multiple sclerosis
NADPH	nicotinamide adenine dinucleotide phosphate
NF	nuclear factor
NFT	neurofibrillary tangle
OPC	oligodendrocyte progenitor cell
PD	Parkinson's disease
PDGF	platelet-derived growth factor
PRR	pattern recognition receptors
RAGE	receptors for advanced glycation end products
RANTES	regulated on activation normal T cell expressed and secreted
ROS	reactive oxygen species
SOD1	superoxide dismutase 1
SNARE	soluble N-ethylmaleimide-sensitive factor attachment protein receptor
TGF-β	transforming growth factor-β
TNF-α	tumor necrosis factor-α
VCAM-1	vascular cell adhesion molecule 1

VEGF	vascular endothelial growth factor
ZO	zonula occluden

1. INTRODUCTION

The central nervous system (CNS) is composed of two main types of cells: neurons and glia. Neurons are electrically excitable cells that respond to a variety of stimuli and participate in synaptic transmission and information processing (Sincich et al., 2009). Glial cells, including astrocytes, microglia and oligodendrocytes, are involved in the homeostatic regulation of the brain functions by providing structural and chemical support to neurons (Jessen, 2004; Kettenmann and Verkhratsky, 2008). Microglia act as the primary immune cells of the brain and are able to respond to a variety of noxious stimuli, including toxins, injury and foreign invaders (Lull and Block, 2010). Oligodendrocytes are involved in the formation of myelin sheath, which surrounds neuronal axons and increases the rate of electrical transmission (Jessen, 2004). Astrocytes, the most prevalent cell type in the brain, are star-shaped cells possessing extended processes, which assist in monitoring homeostatic conditions and metabolic processes in the brain (Jessen, 2004; Meyer and Kaspar, 2016). Specifically, astrocytes have been shown to maintain optimal extracellular conditions through the scavenging of harmful reactive oxygen species (ROS) and by controlling levels of potassium (K^+) in the brain (Kettenmann and Verkhratsky, 2008; Osborn et al., 2016). Astrocytes also contribute to immune responses in the CNS and are involved in synaptic functioning and neuronal repair (Eddleston and Mucke, 1993; Meyer and Kaspar, 2016). Astrocytes support proficient neuronal functioning by supplying neurons with energy substrates (Sherwood et al., 2006), controlling the formation and maintenance of synapses (Nedergaard et al., 2003; Kettenmann and Verkhratsky, 2008), and regulating the uptake of glutamate (Eddleston and Mucke, 1993; Sherwood et al., 2006), thereby moderating neuronal exposure to potentially excitotoxic compounds. Furthermore, astrocytes play a crucial role in the maintenance of the blood-brain barrier (BBB). Astrocytes are able to use their long processes to interact with cells forming the BBB, therefore controlling the flow of water, electrolytes and macromolecules into the CNS; thus, ultimately regulating the extracellular volume and composition in the brain (Nedergaard et al., 2003; Kettenmann and Verkhratsky, 2008).

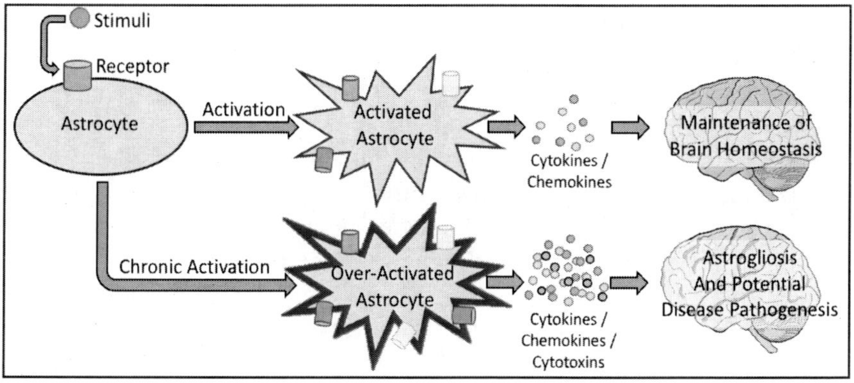

Figure 1. The Activation and Over-Activation of Astrocytes. Astrocytes become activated by a variety of stimuli, which causes changes in astrocyte morphology, receptor expression and secretory profile. This physiological response is necessary for the maintenance of brain health. Over-activation of astrocytes, also termed astrogliosis, occurs when astrocytes are chronically activated causing excessive release of inflammatory molecules, including cytokines, chemokines and cytotoxins. Astrogliosis can also lead to astrocyte atrophy, initiation of chronic inflammation and pathogenesis of various neurodegenerative diseases.

Astrocytes are able to respond to and become activated by a wide variety of stimuli, including the appearance of abnormal molecules in the extracellular space, the absence of signaling molecules that should normally be present and an increase in concentration of various intercellular signaling molecules, such as cytokines and chemokines (Ben Haim et al., 2015). Astrocyte activation is a graded response that occurs rapidly following the disruption of homeostatic conditions in the brain (Ben Haim et al., 2015; Lee and MacLean, 2015). Chronic activation of astrocytes can lead to their reactivity, also known as astrogliosis, which is a more prolonged response that occurs due to astrocytes detecting pathological conditions in the CNS (Eddleston and Mucke, 1993; Lee and MacLean, 2015). Astrogliosis is associated with morphological, transcriptional and secretome changes, which lead to modified astrocytic functions (Ben Haim et al., 2015). Some of these alterations include hypertrophy of the astrocyte cell body, activation of internal signaling cascades, overexpression of the intermediate filament glial fibrillary acidic protein (GFAP) and increased release of pro-inflammatory cytokines (Eddleston and Mucke, 1993; Lee and MacLean, 2015). Under normal physiological conditions, activation of astrocytes can be advantageous, as it can help protect the CNS from pathogens or injury; however, if this activation is sustained for prolonged periods, it can have detrimental effects on surrounding cells of the CNS (Figure 1).

The role of astrocytes in brain metabolism and neurotransmission has been well established; however, recent data indicate that these cells are also critical for the support and maintenance of CNS immune status (Nedergaard et al., 2003; Jessen, 2004). This chapter focuses on neuroimmune functions of astrocytes and outlines the various immune interactions astrocytes have with other cell types of the CNS, including endothelial cells (ECs), neurons, microglia and oligodendrocytes. Furthermore, we discuss the relevance of these interactions in the development and progression of various neurodegenerative diseases.

2. Immune Interactions between Astrocytes and Endothelial Cells

The BBB is a complicated and dynamic neurovascular structure composed of four main elements: ECs, pericytes, secreted basal lamina and astrocytic feet (Abbott et al., 2006). The ECs line blood vessels and enclose the plasma, providing an interface between the blood and the interstitial fluid of the brain. Beyond the ECs lays the basal lamina, a layer of extracellular matrix secreted by ECs, which acts as an anchoring and signaling site for interactions between cells composing the BBB. Pericytes are contractile cells within the basal lamina that help regulate CNS homeostasis by signaling to ECs and astrocytes through direct physical contact as well as paracrine signals (Abbott et al., 2006). Astrocytes provide an additional layer separating the periphery and the brain by extending their feet towards the ECs, surrounding and encasing the basal lamina and pericytes. These cellular units work together to regulate passage of electrolytes, signaling molecules, nutrients and waste between the peripheral circulatory system and the CNS.

Astrocytes, in particular, perform a number of homeostatic functions in regulating the BBB. For example, astrocytes provide physical support to the ECs at the BBB and participate in regulation of blood flow to the brain. They remove neurotoxic molecules from the CNS and 'dump' them into the peripheral bloodstream. Astrocytes maintain osmoregulation of the CNS, regulate tight junctions of ECs, and control passage of molecules across the BBB.

2.1. Interactions between Astrocytes and Endothelial Cells Regulate Passage of Molecules across the Blood Brain Barrier

Astrocytes of the BBB are in constant communication with ECs. One of the ways by which astrocytes regulate homeostasis of the CNS is modulation of the tight junctions between ECs along the BBB. Several key tight junction proteins participate in joining ECs; they include claudins and occludins, junctional adhesion molecules, endothelial selective adhesion molecules, zonula occluden (ZO) proteins and vascular endothelial cadherins (Woodman et al., 1999; Stamatovic et al., 2005; Abbott et al., 2006; Paul et al., 2014). In addition to regulating the BBB and passage of molecules in and out of the brain, astrocytes influence differentiation of ECs during development. Using an *in vitro* co-culture system, Schiera et al., (2003) demonstrated that astrocytes induced differentiation of rodent ECs and stimulated the synthesis of occludin, an important tight junction protein, by ECs (Schiera et al., 2003).

A study by Paul et al., (2014) demonstrated that astrocytic secretion of the inflammatory mediator monocyte chemotactic protein (MCP)-1 down-regulated distribution of claudin-5 along EC tight junctions (Paul et al., 2014), demonstrating that during inflammatory events in the CNS, astrocytes are capable of increasing permeability of the BBB. These findings have been confirmed by other *in vitro* murine cell culture and *in vivo* MCP-1 knockout experiments, which showed that astrocyte-derived MCP-1 enhanced permeability of the BBB 17-fold by means of reduced occludin, claudin-5, ZO-1 and ZO-2 protein expression (Stamatovic et al., 2005). Other studies have demonstrated that during inflammatory events involving viral proteins, astrocytes increase the expression of vascular cell adhesion molecule 1 (VCAM-1) and intercellular adhesion molecule 1 (ICAM-1, CD54), indicating that astrocytes may aid in cellular extravasation across the BBB during pathogenic events (Woodman et al., 1999).

Astrocytes can also positively regulate the expression of tight junction proteins in the BBB. Willis et al., (2004) demonstrated that the lack of astrocytes leads to a remarkable down-regulation in the tight junction proteins claudin-5 and ZO-1 along the BBB in rats (Willis et al., 2004). Not surprisingly, this loss of tight junction proteins was correlated with an increased passage of molecules across the BBB. Furthermore, when astrocyte numbers were restored, EC expression of claudin-5 and ZO-1 was also reinstated (Willis et al., 2004). A study by Gardner et al., (1997) supported these data by showing that astrocyte-conditioned cell culture medium increased the expression of ZO-1 1.7-fold in cultured ECs (Gardner et al.,

1997). However, these observations are not without controversy, as others have failed to reproduce these findings using a different *in vitro* BBB model (Hamm et al., 2004). Modification of the BBB permeability by astrocytes allows for increased passage of molecules, such as cytokines and other immune mediators, to and from the brain (Banks et al., 1995; Banks and Erickson, 2010). In this way, the interaction between astrocytes and ECs of the BBB can have a significant effect on the neuroimmune status of the CNS.

In addition to controlling tight junctions, astrocytes also regulate the passage of nutrients across the BBB by modifying EC expression of receptors and transporters. Mice studies have revealed that under the conditions of hypoxia and starvation, astrocytes upregulate the expression of glucose transporter (GLUT) 1 in ECs of the BBB (Kuchler-Bopp et al., 1999; Boado and Pardridge, 2002). GLUT1 mediates the movement of glucose from the blood to the brain, where glucose acts as the primary source of energy for neurons and glia (Morgello et al., 1995). Won et al., (2012) demonstrated that recurrent moderate hypoglycemia can lead to a 4- or 5-fold increase in microglial activation in non-diabetic and diabetic mice, respectively (Won et al., 2012), suggesting that appropriate glucose levels are required for maintaining glial homeostasis (Jha et al., 2015). Studies have also shown that reduced glucose metabolism is indicative of a neuroinflammatory environment, particularly in AD (Samuraki et al., 2015), and that reduction in glucose transport to the brain via GLUT1 is associated with seizure disorders (Thouin and Crompton, 2016).

Cell culture studies showed that exposure of ECV304 human ECs to cell culture media from C6 rodent astrocytes induced ECV304 cells to express the L-system amino acid transporter (LAT) (Chishty et al., 2002). The LAT carries out transport of large neutral amino acids, such as L-glutamine, L-asparagine, D-leucine and D-phenylalanine from the extracellular matrix to the cytoplasm of the cell (Yanagida et al., 2001; Kim et al., 2002). BBB co-culture model systems support the findings that exposure to astrocyte-conditioned media leads to upregulated expression of LATs in mouse ECs, and enhances the uptake of amino acids L-leucine and L-phenylalanine, but not L-alanine, by ECs (Omidi et al., 2008). Deficiency of amino acids is known to impair immune cell function and increase susceptibility to infection (Li et al., 2007). Therefore, by transporting nutrients across the BBB, astrocytes help maintain brain homeostasis and prevent neuroinflammation.

In addition to regulating the passage of nutrients across the BBB, astrocytes aid in the management of osmotic balance in the CNS. Aquaporin-4 (AQP4) is a water channel located in the basolateral membrane of ECs forming BBB, and also in astrocytes (Mack and Wolburg, 2013). It has been determined that the distribution of AQP4 in astrocytes is polarized, with enhanced expression near the BBB. The AQP4 channel co-localizes with the K^+ channel Kir4.1 (Mack and Wolburg, 2013). The ECs of the BBB and astrocytes communicate and direct water and K^+ movement in and out of the brain via these channels, demonstrating the role astrocytes play in CNS osmoregulation. Imbalance of salts and fluids in the CNS can lead to swelling of the brain and subsequently neuroinflammation; thus, by regulating water and K^+ balance across the BBB, astrocytes can modify the brain's neuroimmune response (Verkman, 2012).

Astrocytes have also been shown to modify properties of the BBB via indirect mechanisms. For example, astrocytes can regulate cerebral blood flow (CBF) by stimulating both the relaxation and contraction of smooth muscles surrounding brain endothelium (Koehler et al., 2009) through calcium (Ca^{2+})-mediated mechanisms (Stobart and Anderson, 2013). Therefore, by controlling vasodilation and vasoconstriction, BBB astrocytes can regulate local blood flow to the specific brain areas. This is of particular importance because inadequate blood flow and ischemic conditions can profoundly alter brain homeostasis and activate glial cells (Chen et al., 2015). Therefore, adequate regulation of CNS blood flow by astrocytes can ultimately prevent the brain from entering a state of chronic inflammation.

2.2. Relevance of Astrocyte and Endothelial Cell Interactions to Neurodegenerative Disease

Since astrocytes are essential for maintaining brain homeostasis, it is not surprising that dysregulation of astrocyte function has been implicated in the pathogenesis of several brain diseases, including Alzheimer's disease (AD) and Parkinson's disease (PD).

AD is a chronic and progressive neurodegenerative disease that is characterized by a substantial decline in cognitive abilities, extensive memory loss and a change in personality and/or behavior (Nagele et al., 2004). This debilitating disease is believed to result from the widespread loss of synapses and degeneration of neurons, which are caused by the deposition of amyloid β protein (Aβ)-containing extracellular plaques, as well as the formation of intracellular neurofibrillary tangles (NFTs) (Mouser et al., 2006; Di Marco et al., 2015). The accumulation of Aβ plaques occurs following the dysregulated cleavage of beta amyloid precursor protein (βAPP), which results in the overproduction and aggregation of Aβ peptides. The NFTs, on the other hand, are formed following the hyperphosphorylation of tau protein, which is normally involved in the stabilization of neuronal microtubules (Blasko et al., 2004; Monteiro-Cardoso et al., 2015). The presence of these protein aggregates is implicated in the pathogenesis of AD, as they have been demonstrated to have toxic effects on surrounding cells, subsequently resulting in neuronal death (Lian and Zheng, 2016; Osborn et al., 2016).

Despite over one hundred years of research, the mechanisms of AD are still poorly understood; however, there are multiple hypotheses that attempt to describe how this disease is initiated and sustained. Increasing evidence implicates neuroinflammation as a fundamental mechanism in the development and progression of AD (Heneka et al., 2015; Venigalla et al., 2015). The presence of Aβ plaques and NFTs is believed to activate surrounding glial cells, including astrocytes and microglia, subsequently causing excess release of non-specific pro-inflammatory mediators, such as tumor necrosis factor (TNF)-α and interleukin (IL)-1β by glial cells (Lull and Block, 2010; Osborn et al., 2016). This secretion of pro-inflammatory molecules has toxic effects on surrounding neurons and also results in the activation of other glial cells, producing a self-perpetuating cycle of glial activation and neuronal death, and ultimately leading to a chronic neuroinflammatory state (Heneka et al., 2015; Venigalla et al., 2015; Lian and Zheng, 2016).

In the early stages of AD, astrocytes undergo astrogliosis, during which they use their processes to surround and engulf the amyloid plaques, thereby protecting the surrounding neurons from the plaque-induced toxic effects (Osborn et al., 2016). However, as the disease progresses, astrogliosis is sustained for a prolonged period and results in astrocytic atrophy (Heneka et al., 2015). More specifically, Mouser et al., (2006) demonstrated that astrocytic exposure to Aβ resulted in the activation of astrocytes, as well as the stimulation of apoptotic processes within the astrocytes; therefore, hindering their neuroprotective abilities and further contributing to the chronic neuroinflammatory state seen in AD.

Chronic neuroinflammation is also a characteristic feature in PD, which is caused by the degeneration of neurons in select brain areas, including the substantia nigra, as well as the disruption of dopaminergic transmission between neurons of the basal ganglia, both of which play a critical role in regulating movement (Nutt and Wooten, 2005; Cabezas et al., 2014). Manifestations of this disease include tremors at rest, slow and rigid movement, instability, as well as abnormal gait (Nutt and Wooten, 2005;

Cabezas et al., 2014; Wang et al., 2015). PD is believed to be initiated and sustained through a variety of neuropathological mechanisms, including the aggregation of α-synuclein protein, an increase in oxidative stress, a dysregulation of mitochondrial functioning and prolonged glial cell activation (Jenner, 2003; Wang et al., 2015; Blanz and Saftig, 2016).

In the early stages of PD, microglia are responsible for the recognition of and response to aggregated α-synuclein. Following this interaction, the activated microglia produce a variety of pro-inflammatory mediators, which results in astrocytic activation and an amplified pro-inflammatory response (Wang et al., 2015). Astrocytes have been shown to contain α-synuclein-positive inclusion bodies themselves (Gu et al., 2010), further implicating them in the progression of PD. More specifically, Gu et al., (2010) demonstrated that presence of excessive α-synuclein within astrocytes promotes the induction of extreme astrogliosis and neuroinflammation, ultimately resulting in the degeneration of both dopaminergic and motor neurons in mice.

ECs are also implicated in the development of AD, as they have been shown to participate in approximately 25% of the removal of Aβ from the brain (Keaney and Campbell, 2015). ECs express an ATP-binding cassette efflux transporter P-glycoprotein, which is involved in the transport of Aβ across the BBB to the periphery (Cirrito et al., 2005; Keaney and Campbell, 2015). Cirrito et al., (2005) showed that mice treated with an inhibitor of the P-glycoprotein transporter had significantly increased levels of Aβ deposits in the brain. ECs in the brain also express the nicotinamide adenine dinucleotide phosphate (NADPH) oxidase 2 enzyme, which acts as a major source of oxidative stress (Cahill-Smith and Li, 2014). In aging brains, as well as those affected by AD or PD, ECs become a target for oxidative attack (Park et al., 2007; Gates et al., 2009), which decreases their ability to maintain BBB integrity, therefore, increasing the permeability of the BBB to harmful pathogens or toxins and facilitating further neuronal death and neuroinflammation (Ueno et al., 2015).

Astrocytes and ECs constantly interact at the BBB to maintain its integrity (Zhao et al., 2015; Lopes Pinheiro et al., 2016). Astrocytes have been shown to release a variety of effector molecules that function to increase and maintain BBB coherence (Holash et al., 1993; Keaney and Campbell, 2015). During the chronic neuroinflammation seen in AD and PD, astrocytic functioning is hindered, which has a negative effect on EC and BBB functioning, thereby compromising regulation of ion flow from the periphery into the CNS (McCarthy and Kosman, 2015; Zhao et al., 2015). Specifically, astrocytic activation has been shown to dysregulate the BBB in such a way that there is an increase in permeability to iron molecules, which in high concentrations can have detrimental effects on brain cells and result in further neurodegeneration (Thomsen et al., 2015). As such, astrocyte-EC interactions have been implicated in the pathogenesis of neurodegenerative diseases (Lopes Pinheiro et al., 2016). However, despite the extensive information available regarding the various ways in which astrocytes interact with ECs, limited research has been conducted regarding the precise mechanisms of how this communication occurs, and its specific role in various forms of brain disease, including AD and PD.

Further examination of the interaction between astrocytes and ECs is required to fully elucidate its implications in the development and progression of neurodegenerative diseases. Investigation of these interactions could reveal novel therapeutic targets for the prevention and treatment of neurodegenerative diseases, including AD and PD.

3. Immune Interactions between Astrocytes and Neurons

Astrocytes participate in various homeostatic mechanisms in the brain, including the formation and maturation of neuronal synapses, synaptic plasticity and synaptic turnover (Chung et al., 2015; Lee and MacLean, 2015). The following section will describe how astrocytes and neurons interact through neuroimmune mechanisms under physiological conditions, and how compromised astrocyte-neuron interactions may contribute to the development of various neurodegenerative diseases.

3.1. Astrocytes Modulate Formation, Plasticity and Elimination of Neuronal Synapses

Astrocytes are essential for synapse formation and maturation, which has been demonstrated through various cell culture experiments; first, in rat retinal ganglion cells (Ullian et al., 2001), and more recently in several other neuron types, including astrocyte interaction with neurons from human pluripotent stem cells (Krencik et al., 2011; Clarke and Barres, 2013). Astrocytes participate in the synaptogenesis of both excitatory glutamatergic synapses and inhibitory gamma-aminobutyric acid (GABA) synapses (Hughes et al., 2010). While the mechanisms of inhibitory synapse formation have yet to be elucidated, the key molecules involved in excitatory synaptogenesis that are secreted by astrocytes include cholesterol, thrombospondin, hevin and glypican (Clarke and Barres, 2013). The secretion of cholesterol from astrocytes is required for the initial dendrite differentiation, as well as pre-synaptic formation (Goritz et al., 2005; Clarke and Barres, 2013). Disturbances in cholesterol secretion by astrocytes are implicated in the pathogenesis of neurodegenerative disorders, such as Huntington's disease (HD) (Valenza et al., 2015).

Another important astrocytic mediator, thrombospondin, interacts with receptors on pre-synaptic and post-synaptic terminals, causing an increase in the formation of synapses (Christopherson et al., 2005). Thrombospondin can interact with integrins, neuroligins and a Ca^{2+} channel subunit (Clarke and Barres, 2013). For instance, the interaction between thrombospondin 1 and neuroligin 1 has been shown to accelerate synapse formation in rat hippocampal neurons, though it was suggested that this effect may be specific to this particular type of neurons (Xu et al., 2010). Hevin is a similar protein that is able to induce synapse formation, as well as regulate presynaptic terminal size (Kucukdereli et al., 2011). Surprisingly, thrombospondin and hevin do not provide an increase in synaptic activity despite their functions in synaptogenesis. These mediators were found to cause an increase in the number of excitatory synapses that were post-synaptically silent, and thus unable to respond to neurotransmitter signals received across the synaptic cleft (Christopherson et al., 2005; Kucukdereli et al., 2011).

The formation and maturation of functional excitatory synapses requires multiple steps and mediators. After induction of synapse formation and synapse increase during development, synapse activity must be stimulated for neurons to function properly. Thus, astrocyte-secreted glypicans were shown to recruit α-amino-3-hydroxy-5-methyl-4-isoxazolepropionic acid receptors (AMPARs) to the synapse, thereby facilitating post-synaptic activity (Allen et al., 2012). It is well established that the regulation of AMPARs is essential for glutamatergic transmission, and the over-excitation of these receptors has severe repercussions on neuron viability (Cebers et al., 1997). Due to the critical role astrocytes play in the complex process of synaptic formation and maturation, their impaired function may lead to deleterious excitotoxic effects on neurons (Ullian et al., 2001).

Astrocytes also help regulate synaptic plasticity, which is characterized by long-term potentiation (LTP) and long-term depression (LTD). Astrocyte-secreted thrombospondin was effective in promoting a form of LTD, called muting, which decreased pre-synaptic neurotransmission (Crawford et al., 2012). More specifically, Hogins et al., (2011) used hypoxic rat hippocampal neurons to demonstrate that muting is neuroprotective against excitotoxic neuronal death (Hogins et al., 2011).

Alternatively, plasticity resulting in the upregulation of excitatory glutamate receptors AMPARs on the post-synaptic terminal results in the increased ability to receive signals, or elevated sensitivity to neurotransmission (Clarke and Barres, 2013). TNF-α, a pro-inflammatory cytokine secreted by astrocytes, was suggested to be important in synaptic plasticity, as it has been shown to recruit AMPARs to the synapse, sensitize synapses to chemical signals and cause an increase in post-synaptic activity (Beattie et al., 2002). As previously mentioned, if AMPAR recruitment is improperly regulated, it has the potential to promote excitotoxicity and neuronal death (Cebers et al., 1997). TNF-α can also be excitotoxic by promoting the secretion of glutamate by astrocytes (Bezzi et al., 2001). Similarly, prostaglandin mediators have been shown to increase astrocytic glutamate secretion (Bezzi et al., 1998). Thus, in a pro-inflammatory environment, these mediators may trigger increased levels of glutamate secretion by astrocytes, and induce an excitotoxic state. Astrocytes are also involved in the removal of glutamate from the synaptic cleft, which demonstrates the complex interface in astrocyte-neuron communication and that the regulation of extracellular glutamate levels is critical for normal CNS function (Wallach et al., 2014). Impairment of this regulation leading to increased extracellular glutamate levels may cause excitotoxicity and neuronal death through the over-stimulation of AMPARs (Cebers et al., 1997).

Astrocytes also participate in synaptic pruning through the phagocytosis of low activity synapses. More specifically, Chung et al., (2013) demonstrated that synapses requiring elimination present phosphatidylserine, which alerts astrocytes to participate in synaptic elimination through the activation of the multiple EGF-l like domains 10 (MEGF10) and MER proto-oncogene tyrosine kinase (MERTK) pathways. Protein S is a circulating extracellular mediator, which regulates astrocyte phagocytosis (Chung et al., 2013). Protein S also functions to maintain the integrity of the BBB, and to protect neurons from hypoxia (Liu et al., 2003; Zhu et al., 2010). Dysregulation of astrocytic phagocytosis could lead to a reduction in the neuroimmune function of astrocytes related to synaptic pruning (Chung et al., 2013).

Synapse elimination by astrocytes may utilize multiple pathways, as a study on the mouse transcriptome determined that astrocytes highly expressed constituents within the MEGF10 pathway, and to a lower extent the MERTK and Crk/Dock1/Elmo pathways (Cahoy et al., 2008). The latter pathway was also implicated in synaptic elimination through astrocytic phagocytosis in *Drosophila* (Tasdemir-Yilmaz and Freeman, 2014). In addition to synapse elimination, resting astrocytes play a role in removal of tissue debris and dead cells by phagocytosis. Primary mouse astrocyte culture experiments showed that this critical function of astrocytes was regulated by the purinergic P2X7 receptors (Yamamoto et al., 2013). Gu et al., (2011) demonstrated that stimulation of P2X7 receptors by ATP resulted in decreased phagocytic activity in human monocytes, therefore suggesting that ATP may be an important signaling molecule mediating the neuroimmune effect of synaptic elimination by astrocytes (Yamamoto et al., 2013). These findings highlight an important neuroimmune role of astrocytes in synaptic elimination and clearance of debris by means of phagocytosis.

Additionally, astrocytes and neurons interact to regulate CBF, which provides the cells with important nutrients and oxygen. In their recent review Lecrux and Hamel (2011) suggested that the most likely mediators for CBF regulation is the release of nitric oxide by neurons, as well as neuronal regulation of astrocytic Ca^{2+} levels, both of which are thought to be induced through neuronal activation by glutamate. Since CBF is critical for oxygen and nutrient supply to neurons, dysregulation of its astrocytic control and reduction in CBF may lead to neuronal death and subsequent development of neurodegenerative conditions (Farkas and Luiten, 2001).

3.2. Relevance of Astrocyte and Neuron Interactions to Neurodegenerative Disease

Astrocytes play an essential role in regulating and providing trophic support for neurotransmission and synaptic development; however, several of these regulatory and support functions are impaired in states of astrogliosis (Avila-Muñoz and Arias, 2014). Neuronal insults resulting from a lack of trophic and metabolic support are exacerbated by dysregulated inflammatory response of astrocytes to injury and disease (Brambilla et al., 2013; Avila-Muñoz and Arias, 2014). In the following section, we will discuss the growing evidence that reduction in the trophic support by astrocytes, as well as their dysregulated inflammatory response, contributes to the pathogenesis of various neurodegenerative diseases including AD, PD, HD and amyotrophic lateral sclerosis (ALS) (Verkhratsky et al., 2014).

3.2.1. Alzheimer's Disease

Studies on AD pathogenesis have illustrated that astrocytes can initiate neuroprotective immune responses directed at clearing pathogenic protein aggregates (Fu and Jhamandas, 2014). Interactions of astrocytes with these aggregates may contribute to AD pathogenesis through deficient neuronal support mechanisms and direct secretion of neurotoxic inflammatory mediators (Verkhratsky et al., 2014).

In AD brains, astrocytes display neuroprotective effects by internalizing and clearing Aβ plaques through an apolipoprotein E-dependent mechanism facilitated by class A scavenger receptors located on their plasma membranes (Fu and Jhamandas, 2014). While this activity may appear beneficial since it can reduce amyloid plaque load, the intracellular accumulation of Aβ alters astrocytic glucose metabolism (Fu and Jhamandas, 2014). This may have a profound effect on AD pathology, as neurons are dependent on astrocytic glucose metabolism for their supply of substrates, including glutathione and lactate. When neurons face diminished levels of these critical substrates, they become increasingly vulnerable to toxic insults and have weakened synaptic capabilities, which adversely affect both transmission and plasticity (Engel, 2014; Verkhratsky et al., 2014). It has also been observed that mid-stage AD brains are characterized by decreased numbers of astrocytes, which may contribute to the progression of the disease through deficient supply of trophic and metabolic factors (Ekonomou et al., 2015).

In late-stage AD, astrocytes may have a detrimental effect on surrounding neurons through the excessive release of pro-inflammatory mediators (Ekonomou et al., 2015). It has been identified that the inflammatory mediator interferon (IFN)-γ in combination with TNF-α or IL-1β may contribute to the generation of Aβ plaques through the induction of β- and γ-secretases (Avila-Muñoz and Arias, 2014). Furthermore, interactions of astrocytes with Aβ induce detrimental signaling cascades including the activation of the transcription factor nuclear factor (NF)-κB, release of complement component 3, dysregulated Ca^{2+} signaling and stimulation of NADPH oxidase (Angelova and Abramov, 2014; Lian et al., 2015; Yates, 2015). Reactive astrocytes have been shown to be the best correlate to cognitive decline in AD brains, as they appear before Aβ plaque deposition and NFT formation (Phillips et al., 2014).

3.2.2. Parkinson's Disease

Contrary to AD, astrogliosis in PD is generally mild to moderate and rarely severe (Verkhratsky et al., 2014). In fact, it has been suggested that lower astrocyte density in PD brains and deficiencies in astrocyte-derived trophic factors are the main mechanisms through which astrocytes contribute to the PD pathogenesis (Panaro and Cianciulli, 2012; L'Episcopo et al., 2014). Under physiological conditions, astrocytes secrete several neurotrophic factors such as glial-derived neurotrophic factor (GDNF), brain-derived neurotrophic factor (BDNF) and ciliary neurotrophic factor (CNTF), all of which have been demonstrated to be expressed at decreased levels in PD (Chung et al., 2010). In fact, transplantation of a unique class of astrocytes characterized by upregulated GDNF, BDNF, neurturin and insulin-like growth factor (IGF)-1 into the striatum of a rat model of PD promoted behavioral recovery (Proschel et al., 2014). Other neuroprotective, neurotrophic and neurogenic factors that are supplied by astrocytes and are expressed abnormally in PD brains include Wnt, nuclear factor erythroid 2-related factor 2 and antioxidants (L'Episcopo et al., 2014). Decline in astrocyte-derived Wnt can contribute to impaired neurogenic potential since this factor has been shown to increase the dopaminergic differentiation potential of neural progenitor cells (L'Episcopo et al., 2014). Activation of the redox-sensitive transcription factor nuclear factor erythroid 2-related factor 2 in astrocytes has been shown to be neuroprotective against α-synuclein toxicity by promoting its degradation through the autophagy-lysosome pathway (Gan et al., 2012). Lastly, induction of astrocyte proliferation and increased antioxidant capacity by 5-hydroxytryptamine protects dopaminergic neurons from oxidative stress (Miyazaki et al., 2013; Mullett et al., 2013).

Despite the neuroprotective effects of astrocytes in PD, it has also been identified that astrocyte-derived pro-inflammatory mediators may contribute to neurodegeneration. For example, reactive astrocytes overexpress monoamine oxidase-B (MAO-B), which has been implicated in the generation of neurotoxins (Ekblom et al., 1993). More specifically, MAO-B converts the environmental factor 1-methyl-4-phenyl-1,2,3,6-tetrahydropyridine (MPTP) to 1-methyl-4-phenylpyridinium (MPP^+), a neurotoxin shown to induce PD through the selective destruction of dopaminergic neurons (Siddiqui et al., 2011). Overall, the contributions of astrocytes to PD pathogenesis are mainly due to deficiencies in trophic support to neurons; however, astrogliosis may, in some cases, directly cause the degeneration of dopaminergic neurons (Lazzarini et al., 2013).

3.2.3. Huntington's Disease

HD is characterized by abnormally long glutamine repeats in the gene encoding the protein huntingtin (Chan and Surmeier, 2014). It has been proposed that the role of astrocytes in HD is not associated with classical astrogliosis; rather, it is characterized by the loss of normal astrocytic function or gain of abnormal astrocyte functions as a result of mutant huntingtin (mHTT) gene expression (Gunčová et al., 2011; Chan and Surmeier, 2014; Tong et al., 2014). Abnormalities have been observed in astrocytes with 144 glutamine repeats, which displayed defective glutamate uptake leading to increased striatal extracellular glutamate levels and increased excitability of neurons (Verkhratsky et al., 2014). Such glutamate uptake deficiency could result from decreased expression of the glutamate aspartate transporter and the glutamate transporter 1 (GLT-1) (Faideau et al., 2010). Additionally, it has been identified that HD astrocytes exhibit enhanced Ca^{2+}-dependent exocytosis of glutamate, augmenting extracellular glutamate levels and contributing to the early-stage progression of HD (Faideau et al., 2010; Verkhratsky et al., 2014).

In addition to disrupting extracellular glutamate regulation, mHTT reduces the capacity of astrocytes to function in synaptic homeostasis mechanisms. Specifically, astrocytes with mHTT have an impaired ability to regulate extracellular levels of GABA, cationic salts, as well as other substances including cholesterol and ascorbic acid. The capacity of depolarized astrocytes to provide GABA for tonic inhibition is strongly reduced in HD (Wójtowicz et al., 2013). As the principal inhibitory neurotransmitter, GABA is critical for controlling neuronal excitability. Low GABA tone, in conjunction with enhanced glutamate levels, has been show to facilitate the increased excitability of neurons in HD. Furthermore, HD astrocytes are deficient in functional Kir4.1 K^+ channels, resulting in elevated extracellular levels of K^+, which could lead to increased medium spiny neuron excitability (Chan and Surmeier, 2014; Khakh and Sofroniew, 2014; Tong et al., 2014).

Neurons depend on astrocytes for the supply of cholesterol. In HD, severely reduced cholesterol biosynthesis proves detrimental to neurons as evidenced by reduced neurite outgrowth and synaptic activity (Valenza et al., 2015). Lastly, it has been identified that in the presymptomatic stages of HD, there is abnormal flux of the antioxidant ascorbic acid from astrocytes to neurons (Acuña et al., 2013). The same study identified that impairment of ascorbic acid uptake could contribute to the early metabolic failure of HD neurons leading to their death.

While classical astrogliosis is not considered to be a contributing factor to HD pathogenesis, several studies have identified that mHTT does have an effect on the inflammatory state of astrocytes (Gunčová et al., 2011; Chan and Surmeier, 2014; Tong et al., 2014). More specifically, several signaling cascades related to the induction and propagation of the translation and activity of NF-κB are affected in mHTT astrocytes. Firstly, it has been demonstrated that mHTT hinders activation of the neurotrophic factor chemokine (C-C motif) ligand 5 (CCL5), also known as regulated on activation normal T cell expressed and secreted (RANTES), promoter in HD astrocytes by reducing the availability of astrocytic NF-κB-p65 (Chou et al., 2008). Secondly, HD astrocytes exhibit elevated IκB kinase (IKK) activity, which propagates NF-κB activation and results in an upregulation of pro-inflammatory factors (Hsiao et al., 2013). Several studies have demonstrated that pro-inflammatory factors resulting from elevated IKK activity in astrocytes elicit neurotoxic activity, and that blockage of IKK activity is sufficient to ameliorate this neurotoxicity (Hsiao et al., 2013; Boussicault et al., 2014). Lastly, it has been demonstrated that downstream signaling of the receptors for advanced glycation end products (RAGE) is altered in HD astrocytes. RAGE-positive astrocytes had increased translocation of NF-κB to the nucleus, suggesting involvement of astrocytic RAGE in HD pathology (Kim et al., 2015).

3.2.4. Amyotrophic Lateral Sclerosis

ALS is a rapidly progressive, neuromuscular disease affecting both upper and lower motor neurons. Superoxide dismutase 1 (SOD1) is implicated in sporadic ALS pathogenesis and, upon mutation, is sufficient to cause ALS through a toxic gain of function (Pehar et al., 2014). Similar to HD, atrophy and degeneration of astrocytes are associated with a loss of function, which precedes neuronal death and clinical symptoms in the initial stages of ALS (Verkhratsky et al., 2014). Deficient astrocyte function contributes to ALS through dysregulated extracellular neurotransmitter concentrations, ionic and metabolic homeostasis, as well as loss of trophic support for neurons (Thangavelu et al., 2011; Hooten et al., 2015). Astrocytes in ALS have been shown to impair glutamate transport resulting from diminished expression of GLT-1, which leads to excitotoxic neuronal injury (D'Amico et al., 2013; Olmos and Lladó, 2014; Hooten et al., 2015). It has been demonstrated that transplantation of astrocytes with improved glutamate transport capacity may delay both motor neuron loss and the decline in motor function seen in mutant SOD1 ALS animal models (Malaspina et al., 2015).

Examination of astrocyte-mediated ionic homeostasis revealed that mutant SOD1 astrocytes have elevated intracellular Ca^{2+} levels, which cause neurotoxicity in ALS through soluble N-ethylmaleimide-sensitive factor attachment protein receptor (SNARE)-dependent exocytosis mechanisms

(Kawamata et al., 2014). Moreover, motor neurons cultured in mutant SOD1 astrocyte conditioned medium *in vitro* displayed enhanced sodium V channel-mediated excitability and Ca^{2+} influx. These alterations in ionic balance resulted in a greater level of intracellular ROS and subsequent toxicity, mediated at least in part by mitochondrial alterations (Rojas et al., 2015). It has also been identified that astrocytes in the pre-symptomatic stages of ALS have reduced lactate metabolism that contributes to disease progression through a loss of neuronal support function, rather than adoption of a toxic phenotype by astrocytes (Baker et al., 2015). Lastly, it has been identified that uninterrupted trophic input by astrocyte-derived factors such as vascular endothelial growth factor (VEGF), GDNF, CNTF and IGF-1 is required to sustain normal motor neuron functioning (Tovar-Y-Romo et al., 2014).

Similar to their role in other neurodegenerative diseases, astrocytes in ALS contribute to the neuroimmune response in several ways. It has been noted that the adoption of a reactive phenotype by astrocytes predominantly occurs in later stages of ALS and, once this stage is reached, astrocytes may exacerbate neuronal damage and actively mediate neuronal death (Verkhratsky et al., 2014; Baker et al., 2015). Several alterations in the inflammatory responses of astrocytes have been observed in ALS and include increased phagocytic activity, glial-scar formation, dysregulated agonistic activity at the TNF receptor 2, secretion of cytokines and chemokines, as well as elevated expression of cyclooxygenase enzymes (Phani et al., 2012; Verkhratsky et al., 2014; Baker et al., 2015; Tortarolo et al., 2015). Increased phagocytic activity of astrocytes has been implicated in mediating neuronal insult; however, it could also protect neurons from contact-induced cell death through the clearance of extracellular debris (Baker et al., 2015). It is interesting to note that astrocytic phagocytosis of apoptotic or necrotic cell debris does not induce their expression of pro-inflammatory cytokines (Baker et al., 2015). As an additional beneficial effect, astrocytes are able to ameliorate ALS symptoms by physically isolating injured CNS areas through the formation of glial scars (Phani et al., 2012).

In contrast to the above beneficial effects, astrocytes have also been implicated in detrimental immune activity, which may cause neuronal death. TNF receptor 2 expressed on astrocytes and motor neurons is implicated in motor neuron loss in both mutant SOD1 primary cell co-cultures and mice models (Tortarolo et al., 2015). Furthermore, reactive astrocytes release factors that can directly induce neuronal apoptosis through the p75 or Bax pathways (Phani et al., 2012). Other potentially cytotoxic factors secreted by reactive astrocytes include various pro-inflammatory cytokines and products of arachidonic acid metabolism, including leukotrienes and prostaglandins (Phani et al., 2012; Verkhratsky et al., 2014). Specifically, reactive astrocytes produce elevated levels of prostaglandin PGE_2, which is a well-known pro-inflammatory mediator and potent inducer of astrocytic glutamate secretion (Verkhratsky et al., 2014).

4. Immune Interactions between Astrocytes and Microglia

Microglia are the resident macrophages of the brain, which are considered professional immune cells due to their myeloid lineage, as well as their innate ability to detect and clear pathogens and damaged cells through phagocytosis (Chen et al., 2015). Microglia continuously monitor the brain microenvironment and use pattern recognition receptors (PRRs) to recognize structurally conserved molecules on the surface of various microorganisms (Shinozaki et al., 2014). Therefore, microglia are able to detect invading pathogens, as well as the appearance of abnormal disease-related proteins and molecular structures (Shinozaki et al., 2014). Under physiological conditions, microglia regulate neuronal activity, synaptic connectivity and programmed cell death (Madry and Attwell, 2015). Following interaction with harmful pathogens or abnormal molecules, microglia become activated and produce a variety of pro-inflammatory cytokines and ROS, aimed at eliminating the noxious stimuli (Escamilla-Tilch et al., 2013).

In addition to their pivotal role in immune surveillance, microglia also facilitate the immune responses within the brain (Norden et al., 2014). The acute inflammatory responses orchestrated by microglia are generally beneficial to the CNS, contributing to the removal of toxic substances and subsequent repair of damaged tissues. However, this initial positive neuroinflammatory response can turn into detrimental prolonged and chronic activation (Polazzi and Monti, 2010). Such reactive microglia can cause neuronal injury and death through the production and secretion of neurotoxic factors and inflammatory mediators, which subsequently engage neighboring cells, including astrocytes, and establish a vicious cycle of autocrine and paracrine amplification of inflammation and neurodegeneration (Niranjan, 2014). Microglia also directly communicate with astrocytes by a variety of signaling molecules, including pro-inflammatory cytokines, chemokines and ATP.

Astrocytes produce regulatory factors that influence CNS immunity by controlling activated microglia (Norden et al., 2014). Activated astrocytes can produce cytokines, which enhance microglial activation and act as paracrine factors maintaining the chronically reactive state of all glial cells (Ben Haim et al., 2015). Microglia, in turn, can indirectly alter excitatory neurotransmission by communicating with astrocytes through ATP-dependent mechanisms (Madry and Attwell, 2015). High levels of extracellular ATP released by degenerating cells have been found to bind to astrocytic P2X receptors and induce further release of ATP. Through this pathway, astrocytes can amplify the local increase in ATP concentration, which has been shown to facilitate microglial chemotaxis following focal brain damage (Madry and Attwell, 2015). Activation of microglial P2Y1 receptors by elevated ATP concentrations can also induce extension of microglial processes (Haynes et al., 2006; Chen et al., 2015).

Microglial secretion of ATP has recently been shown to activate the P2Y1 receptors of astrocytes and trigger ATP release by astrocytes, which in turn upregulates synaptic transmission (Pascual et al., 2012). P2Y1 receptor activation in astrocytes has also been reported to induce neuroprotective effects against oxidative stress, trauma and neurotoxins (Shinozaki et al., 2014). It has been suggested that microglia-astrocyte interactions are required for bacterial lipopolysaccharide (LPS)-induced expression of pro-inflammatory factors by astrocytes (Chen et al., 2015). Microglia-derived TNF-α plays a pivotal role as a paracrine signal, which is used to regulate the neuroprotective functions of astrocytes. Taken together, these findings indicate that astrocytes may not be able to recognize the innate immune stimuli such as LPS directly, but rather depend on crosstalk with microglia to elicit release of neurotrophic factors (Chen et al., 2015).

Transforming growth factor (TGF)-β and IL-10 are two key anti-inflammatory cytokines that demonstrate neuroprotective activity within the CNS by modulating astrocyte and microglial activation (Norden et al., 2014). Numerous studies have demonstrated that IL-10 has anti-inflammatory effects on microglia by inducing astrocytic production of TGF-β which, in turn, inhibits the production of pro-inflammatory molecules such as nitric oxide by activated microglia (Norden et al., 2014). Thus, the interaction between astrocytes and microglia through secreted immune mediators could assist in regulating neuroinflammation.

Modulation of number of synapses and their function is emerging as a critical component of the role glia play in the process of neurodegeneration (Jebelli et al., 2015). Microglia in a neurocytopathic state are characterized by an enhanced release of inflammatory cytokines and chemokines, as well as reactive oxygen and nitrogen species, resulting in elimination of synapses and the excessive neuronal death seen during neurodegeneration (Akiyama et al., 2000; Youssef et al., 2010; Hashioka et al., 2011). Once activated by pro-inflammatory mediators, both microglia and astrocytes can release mediators capable of further amplification of glial responses. For example, purines and pro-inflammatory cytokines may be released by reactive astrocytes and, in larger amounts, by activated microglia (Ben Haim et al., 2015). Microglial activation is a common feature found at the early stages of various CNS disorders, during which microglia migrate toward the source of trauma, phagocytose debris and initiate inflammatory reactions (Shinozaki et al., 2014). Reactive astrogliosis is also observed in neuroinflammation and serves as a marker for several neuropathologies (Chen et al., 2015). Such glial-driven inflammation is now widely recognized as an important underlying component of a wide range of neurodegenerative diseases including AD, PD and ALS (Jebelli et al., 2015). Novel therapeutic routes for the prevention and treatment of these neurodegenerative diseases may be revealed through further investigation of the interactions that occur between astrocytes and microglia.

5. Immune Interactions between Astrocytes and Oligodendrocytes

Oligodendrocytes are most prominently known for their contributions to the development of the myelin sheath that surrounds the axons of neurons (Liu and Zhou, 2013). Myelin sheath acts as a protective electrical insulator, and increases the speed at which an action potential (AP) travels down an axon (Siegel and Agranoff, 1999). Currently, there is little information on the interaction between oligodendrocytes and astrocytes under physiological conditions. In the absence of oligodendrocyte-related pathologies, oligodendrocyte-astrocyte interactions are most likely unrelated to immune responses. Under normal circumstances, astrocytes have been primarily found to aid in the active turnover of mature oligodendrocytes. Chemokines and bone morphogenetic proteins secreted by astrocytes have also been found to modulate the behaviour of oligodendrocyte progenitor cells (OPCs), and inhibit OPC development, respectively (Clemente et al., 2013). Loss of myelin sheath or demyelination leads to ineffective transmission of APs, as well as axonal damage (Liu and Zhou, 2013).

Multiple sclerosis (MS) is a prevalent neurodegenerative disease characterized by demyelination and chronic inflammation, which are caused by dysfunctional immune regulation (Miller, 2005; Clarner et al., 2011). This results in an autoimmune attack on oligodendrocytes, thereby damaging the myelin sheath around the axon (Miller, 2005; Clarner et al., 2011). The demyelination of axons accounts for the symptoms associated with MS, such as lack of coordination and balance, as well as impaired cognitive function (Miller, 2005; Dabrowski et al., 2015). MS is initiated by the development of focal lesions, which are caused by damage to cerebral or spinal white matter (Brück, 2005; Inglese, 2006). Sclerotic plaques is another name used to describe these lesions, which have been found to cause acute local inflammation characterized by recruitment of reactive glial cells, as well as peripheral immune cells. Chronic inflammation, propagated by glial release of pro-inflammatory cytokines, can cause further demyelination and destruction of oligodendrocytes, as well as axonal damage (Brück, 2005). This section will focus on the contrasting roles astrocytes play in MS by contributing to both inflammation and remyelination.

Activated astrocytes contribute to the chronic inflammation observed in MS through the release of cytokines that recruit microglia and other macrophages from the periphery (Savarin et al., 2015). Specifically, astrocytes in an MS rat model demonstrated an increased secretion of the pro-inflammatory cytokine IL-6, with such release restricted to areas of demyelination. Sustained astrocyte expression and release of IL-6 have been found to increase deterioration of myelin and induce astrogliosis, while anti-IL-6 treatments led to decreased myelin loss and reduced axonal damage (Savarin et al., 2015). Astrocytes can also release molecules that inhibit the repair of demyelination. Hammond et al., (2014) demonstrated that endothelin-1 (ET-1), a signalingmolecule expressed on reactive astrocytes, impedes remyelination and differentiation of OPCs into mature oligodendrocytes (Hammond et al., 2014).

To reverse the effects of demyelination, mature oligodendrocytes derived from OPCs are required. OPCs migrate to sites of injury and differentiate into mature oligodendrocytes that are capable of producing myelin (de Castro et al., 2013). This transformation is accomplished following the expression and release of growth factors, chemoattractants and adhesion molecules from reactive astrocytes, which interact with the OPCs (de Castro et al., 2013). One of the most characterized growth factors secreted by astrocytes is the platelet-derived growth factor (PDGF). The PDGF receptor is present on OPCs, and is often used as a marker for OPCs. The interaction between PDGF and its receptor results in OPC proliferation (Moore et al., 2011). Vana et al., (2008) demonstrated that, following demyelination induced by cuprizone, a copper chelator, the secretion of PDGF from human astrocytes increased the number of oligodendrocytes in lesions (Vana et al., 2007; Torkildsen et al., 2008). The presence of oligodendrocytes in lesions is beneficial and necessary for remyelination.

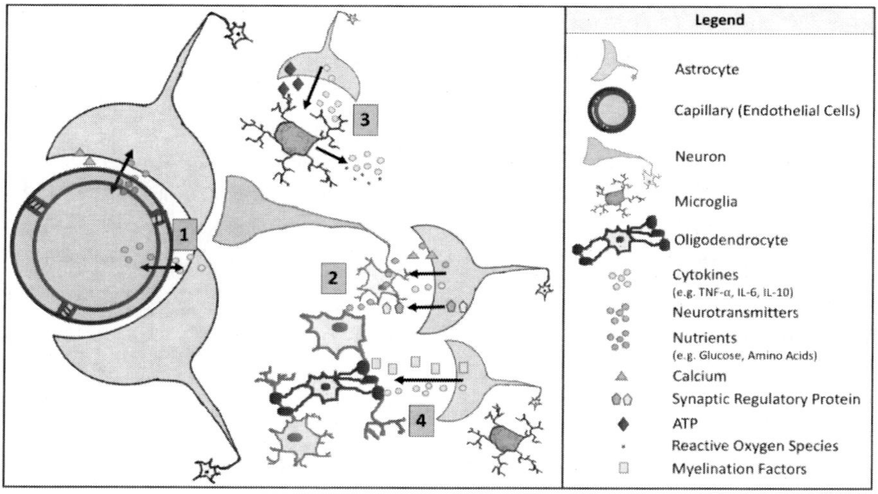

Figure 2. Interactions Between Astrocytes and Other Cell Types of the Central Nervous System. 1) Astrocytes regulate passage of cytokines, nutrients and ions across the BBB, by modulating the permeability of tight junctions, expression of transporters and cerebral blood flow. 2) Astrocytes influence neuronal functions by recycling neurotransmitters and inducing synaptic neurotransmission through Ca^{2+}-mediated mechanisms. Astrocytes facilitate synaptic remodeling by signaling to neurons via thrombospondin, hevin, MEGF10 and MERTK. Astrocytes also communicate with neurons by secreting cytokines during inflammatory events. 3) Astrocytes regulate microglial activity by secreting ATP and cytokines, which can recruit microglia to sites of inflammation via chemotaxis, and activate the microglia to transform into either a pro- or anti-inflammatory state. 4) Astrocytes secrete cytokines and factors, such as ET-1 and PDGF, which affect oligodendrocytes and influence neuronal myelination. These interactions involving astrocytes maintain brain homeostasis and mediate neuroimmune interactions during both normal physiological functioning and pathological events.

Astrocytes also secrete extracellular matrix molecules, such as laminins, which interact with dystroglycan, an extracellular matrix receptor found on

OPCs and mature oligodendrocytes (Colognato et al., 2007; Moore et al., 2011). This interaction facilitates myelin production, OPC differentiation and the survival of oligodendrocytes. Therefore, a decrease in dystroglycan expression is associated with the degradation of myelin (Colognato et al., 2007).

Astrocyte-oligodendrocyte signaling through gap junctions is also important in remyelination. Such junctions allow for intercellular communication between these two types of glial cells. A decrease in gap junction formation occurs in areas of chronic inflammation, which results in abnormal functioning of glial cells and exacerbated MS progression (Markoullis et al., 2014).

The implications of the astrocyte-oligodendrocyte interactions in MS have often been underestimated in the past (Moore et al., 2011). As a result, many of the mechanisms and details regarding the role of glia in MS are still elusive. However, potential therapeutic strategies for the prevention and treatment of MS may be derived from examining the interactions between astrocytes and oligodendrocytes.

CONCLUSION

By interacting with other CNS cell types, astrocytes are essential for the support and maintenance of CNS homeostasis through several mechanisms (Figure 2). By interacting with ECs, astrocytes regulate the integrity of the BBB and participate in the control of osmosis, vasodilation and vasoconstriction, as well as the exchange of nutrients, metabolites and toxins between the CNS and periphery. Moreover, astrocytes are essential for normal synaptic development and function as well as neuronal differentiation and repair. It has been established that astrocytes supply neurons with necessary trophic and metabolic support, control the formation and maintenance of synapses and regulate extracellular concentrations of glutamate. Under inflammatory conditions, astrocytes regulate the activation of microglia and work synergistically with these primary immune cells to protect the CNS from insult. Lastly, astrocytes interact closely with oligodendrocytes to support axonal myelination, and regulate oligodendrocyte differentiation through the gap junction signaling and the secretion of paracrine factors.

Table 1. Astrocytes in Brain Pathologies

Disease	Interactions of Astrocytes with Other Cell Types
Alzheimer's Disease	• AD brains contain reactive astrocytes (Fu and Jhamandas, 2014). • Astrocyte interaction with Aβ leads to activation of the transcription factor NF-κB, release of complement component 3, dysregulated Ca^{2+} signaling, and stimulation of NADPH oxidase (Angelova and Abramov, 2014; Lian et al., 2015; Yates, 2015). • Astrocytes become over-activated, leading to significant cognitive decline (Phillips et al., 2014). • Sustained astrocyte activation results in astrocytic atrophy (Heneka et al., 2015).
Parkinson's Disease	• PD brains contain reactive astrocytes (Verkhratsky et al., 2014). • α-Synuclein-positive inclusion bodies could be present in astrocytes (Gu et al., 2010). • There is a reduction in astrocyte-derived trophic factors, including GDNF, BDNF, CNTF, Wnt and nuclear factor erythroid 2-related factor 2 (Chung et al., 2010; Panaro and Cianciulli, 2012; L'Episcopo et al., 2014).
Huntington's Disease	• Astrocytes in HD brains gain abnormal functions due to accumulation of mHTT (Gunčová et al., 2011; Chan and Surmeier, 2014; Tong et al., 2014). • Astrocytes have elevated IKK activity and NF-κB activation, resulting in increased neuroinflammation (Hsiao et al., 2013). • Astrocytic RAGE expression is upregulated compared to non-HD astrocytes (Kim et al., 2015). • Reduced extracellular glutamate uptake by astrocytes leads to increased excitability of neurons (Verkhratsky et al., 2014). • Glutamate release from astrocytes is increased (Faideau et al., 2010; Verkhratsky et al., 2014). • Astrocytes produce less cholesterol leading to reduced neurite outgrowth and impaired synaptic activity (Valenza et al., 2015).
Amyotrophic Lateral Sclerosis	• In ALS, astrocytic atrophy, degeneration and loss of function precede neuronal death and clinical symptoms (Verkhratsky et al., 2014). • Astrocyte glutamate transport is impaired due to decreased expression of GLT-1, leading to increased excitotoxicity (D'Amico et al., 2013; Olmos and Lladó, 2014; Hooten et al., 2015). • Increased phagocytic activity of astrocytes, enhanced glial-scar formation, upregulated cytokine secretion and elevated expression of cyclooxygenase enzymes by astrocytes have been reported (Phani et al., 2012; Verkhratsky et al., 2014; Baker et al., 2015; Tortarolo et al., 2015).
Multiple Sclerosis	• MS brains contain reactive astrocytes (Savarin et al., 2015). • Astrocytic secretion of IL-6 and ET-1 leads to increased degradation of myelin, impaired remyelination and enhanced MS progression (Hammond et al., 2014). • Astrocyte-mediated inflammation leads to weakening of gap junctions and enhanced progression of MS (Markoullis et al., 2014).

While astrocytes function as essential support cells under normal physiological conditions, it is now widely accepted that astrocytes can adopt unique activation patterns throughout the pathological stages of various neurological disorders (Table 1). Astrocyte activation could be induced by a variety of disease-specific signals, leading to a graded response aimed at protecting the CNS from disease and injury. Astrogliosis develops under the conditions of chronic activation and is characterized by morphological, transcriptional and secretome changes, which have been implicated as contributing factors in the pathogenesis of neurological diseases. Chronic activation of astrocytes has been demonstrated to impair their ability to maintain BBB integrity, thereby increasing the permeability of the BBB and compromising osmotic homeostasis. Moreover, chronically activated astrocytes impair proper neuronal functioning through reduced provision of energy substrates and growth factors, as well as through dysregulated inflammatory response, which can increase the susceptibility of neurons to excitotoxicity. The adverse inflammatory response of astrocytes in pathological conditions is facilitated by interactions with microglia, in which autocrine and paracrine signaling may cause amplified secretion of pro-inflammatory mediators and cytotoxins, leading to a loss of neurons and synapses. Astrogliosis may further contribute to neuropathology through the secretion of factors that enhance demyelination of axons and impede their remyelination. Thus, astrocytes are relevant and actively contribute to the pathogenic mechanisms of neurodegenerative diseases through diverse immune interactions with other key cell types of the CNS. Understanding the patterns of these interactions and intercellular signaling pathways could lead to identification of novel therapeutic targets for the treatment of a wide range of neurodegenerative diseases.

ACKNOWLEDGMENTS

This work was supported by grants from the Natural Sciences and Engineering Research Council of Canada (NSERC) and the Jack Brown and Family Alzheimer's Disease Research Foundation.

REFERENCES

Abbott, N. J., Ronnback, L. & Hansson, E. (2006). "Astrocyte-endothelial interactions at the blood-brain barrier." *Nat Rev Neurosci*, *7*, 41-53.

Acuña, A. I., Esparza, M., Kramm, C., et al., (2013). "A failure in energy metabolism and antioxidant uptake precede symptoms of Huntington's disease in mice." *Nat Commun*, *4*, 2917.

Akiyama, H., Barger, S., Barnum, S., et al., (2000). "Inflammation and Alzheimer's disease." *Neurobiol Aging*, *21*, 383-421.

Allen, N. J., Bennett, M. L., Foo, L. C., et al., (2012). "Astrocyte glypicans 4 and 6 promote formation of excitatory synapses via GluA1 AMPA receptors." *Nature*, *486*, 410-414.

Angelova, P. R. & Abramov, A. Y. (2014). "Interaction of neurons and astrocytes underlies the mechanism of Aβ-induced neurotoxicity." *Biochem Soc Trans*, *42*, 1286-1290.

Avila-Muñoz, E. & Arias, C. (2014). "When astrocytes become harmful: functional and inflammatory responses that contribute to Alzheimer's disease." *Ageing Res Rev*, *18*, 29-40.

Baker, D. J., Blackburn, D. J., Keatinge, M., et al., (2015). "Lysosomal and phagocytic activity is increased in astrocytes during disease progression in the SOD1 G93A mouse model of amyotrophic lateral sclerosis." *Front Cell Neurosci*, *9*, 410.

Banks, W. A. & Erickson, M. A. (2010). "The blood-brain barrier and immune function and dysfunction." *Neurobiol Dis*, *37*, 26-32.

Banks, W. A., Kastin, A. J. & Broadwell, R. D. (1995). "Passage of cytokines across the blood-brain barrier." *Neuroimmunomodulation*, *2*, 241-248.

Beattie, E. C., Stellwagen, D., Morishita, W., et al., (2002). "Control of synaptic strength by glial TNFalpha." *Science*, *295*, 2282-2285.

Ben Haim, L., Carrillo-de Sauvage, M. A., Ceyzeriat, K. & Escartin, C. (2015). "Elusive roles for reactive astrocytes in neurodegenerative diseases." *Front Cell Neurosci*, *9*, 278.

Bezzi, P., Carmignoto, G., Pasti, L., et al., (1998). "Prostaglandins stimulate calcium-dependent glutamate release in astrocytes." *Nature*, *391*, 281-285.

Bezzi, P., Domercq, M., Brambilla, L., et al., (2001). "CXCR4-activated astrocyte glutamate release via TNFalpha: amplification by microglia triggers neurotoxicity." *Nat Neurosci*, *4*, 702-710.

Blanz, J. & Saftig, P. (2016). "Parkinson's disease: acid-glucocerebrosidase activity and alpha-synuclein clearance." *J Neurochem* Epub ahead of print. doi:10.1111/jnc.13517.

Blasko, I., Stampfer-Kountchev, M., Robatscher, P., Veerhuis, R., Eikelenboom, P. & Grubeck-Loebenstein, B. (2004). "How chronic inflammation can affect the brain and support the development of Alzheimer's disease in old age: the role of microglia and astrocytes." *Aging Cell*, *3*, 169-176.

Boado, R. J. & Pardridge, W. M. (2002). "Glucose deprivation and hypoxia increase the expression of the GLUT1 glucose transporter via a specific mRNA cis-acting regulatory element." *J Neurochem*, *80*, 552-554.

Boussicault, L., Hérard, A. S., Calingasan, N., et al., (2014). "Impaired brain energy metabolism in the BACHD mouse model of Huntington's disease: critical role of astrocyte-neuron interactions." *J Cereb Blood Flow Metab*, *34*, 1500-1510.

Brambilla, L., Martorana, F. & Rossi, D. (2013). "Astrocyte signaling and neurodegeneration: New insights into CNS disorders." *Prion*, *7*, 28-36.

Brück, W. (2005). "The pathology of multiple sclerosis is the result of focal inflammatory demyelination with axonal damage." *J Neurol*, *252*, 3-9.

Cabezas, R., Avila, M., Gonzalez, J., et al., (2014). "Astrocytic modulation of blood brain barrier: perspectives on Parkinson's disease." *Front Cell Neurosci*, *8*, 211.

Cahill-Smith, S. & Li, J. M. (2014). "Oxidative stress, redox signalling and endothelial dysfunction in ageing-related neurodegenerative diseases: a role of NADPH oxidase 2." *Br J Clin Pharmacol*, *78*, 441-453.

Cahoy, J. D., Emery, B., Kaushal, A., et al., (2008). "A transcriptome database for astrocytes, neurons, and oligodendrocytes: a new resource for understanding brain development and function." *J Neurosci*, *28*, 264-278.

Cebers, G., Zhivotovsky, B., Ankarcrona, M. & Liljequist, S. (1997). "AMPA neurotoxicity in cultured cerebellar granule neurons: mode of cell death." *Brain Res Bull*, *43*, 393-403.

Chan, C. S. & Surmeier, D. J. (2014). "Astrocytes go awry in Huntington's disease." *Nat Neurosci*, *17*, 641-642.

Chen, S. H., Oyarzabal, E. A., Sung, Y. F., et al., (2015). "Microglial regulation of immunological and neuroprotective functions of astroglia." *Glia*, *63*, 118-131.

Chen, Y. J., Nguyen, H. M., Maezawa, I., et al., (2015). "The potassium channel KCa3.1 constitutes a pharmacological target for neuroinflammation associated with ischemia/reperfusion stroke." *J Cereb Blood Flow Metab* Epub ahead of print. pii: 0271678X15611434.

Chishty, M., Reichel, A., Begley, D. J. & Abbott, N. J. (2002). "Glial induction of blood-brain barrier-like L-system amino acid transport in the ECV304 cell line." *Glia*, *39*, 99-104.

Chou, S. Y., Weng, J. Y., Lai, H. L., et al., (2008). "Expanded-Polyglutamine Huntingtin Protein Suppresses the Secretion and Production of a Chemokine (CCL5/RANTES) by Astrocytes." *J Neurosci*, *28*, 3277-3290.

Christopherson, K. S., Ullian, E. M., Stokes, C. C., et al., (2005). "Thrombospondins are astrocyte-secreted proteins that promote CNS synaptogenesis." *Cell*, *120*, 421-433.

Chung, W. S., Clarke, L. E., Wang, G. X., et al., (2013). "Astrocytes mediate synapse elimination through MEGF10 and MERTK pathways." *Nature*, *504*, 394-400.

Chung, W. S., Welsh, C. A., Barres, B. A. & Stevens, B. (2015). "Do glia drive synaptic and cognitive impairment in disease?" *Nat Neurosci*, *18*, 1539-1545.

Chung, Y. C., Ko, H. W., Bok, E., et al., (2010). "The role of neuroinflammation on the pathogenesis of Parkinson's disease." *BMB Rep*, *43*, 225-232.

Cirrito, J. R., Deane, R., Fagan, A. M., et al., (2005). "P-glycoprotein deficiency at the blood-brain barrier increases amyloid-beta deposition in an Alzheimer disease mouse model." *J Clin Invest*, *115*, 3285-3290.

Clarke, L. E. & Barres, B. A. (2013). "Emerging roles of astrocytes in neural circuit development." *Nat Rev Neurosci*, *14*, 311-321.

Clarner, T., Parabucki, A., Beyer, C. & Kipp, M. (2011). "Corticosteroids impair remyelination in the corpus callosum of cuprizone-treated mice." *J Neuroendocrinol*, *23*, 601-611.

Colognato, H., Galvin, J., Wang, Z., et al., (2007). "Identification of dystroglycan as a second laminin receptor in oligodendrocytes, with a role in myelination." *Development*, *134*, 1723-1736.

Crawford, D. C., Jiang, X., Taylor, A. & Mennerick, S. (2012). "Astrocyte-derived thrombospondins mediate the development of hippocampal presynaptic plasticity *in vitro*." *J Neurosci*, *32*, 13100-13110.

D'Amico, E., Factor-Litvak, P., Santella, R. M. & Mitsumoto, H. (2013). "Clinical perspective on oxidative stress in sporadic amyotrophic lateral sclerosis." *Free Radic Biol Med*, *65*, 509-527.

Dabrowski, W., Kwiecien, J. M., Rola, R., et al., (2015). "Prolonged subdural infusion of kynurenic acid Is associated with dose-dependent myelin damage in the rat spinal cord." *PLoS One*, *10*.

de Castro, F., Bribián, A. & Ortega, M. C. (2013). "Regulation of oligodendrocyte precursor migration during development, in adulthood and in pathology." *Cell Mol Life Sci*, *70*, 4355-4368.

Di Marco, L. Y., Venneri, A., Farkas, E., Evans, P. C., Marzo, A. & Frangi, A. F. (2015). "Vascular dysfunction in the pathogenesis of Alzheimer's disease--A review of endothelium-mediated mechanisms and ensuing vicious circles." *Neurobiol Dis*, *82*, 593-606.

Eddleston, M. & Mucke, L. (1993). "Molecular profile of reactive astrocytes--implications for their role in neurologic disease." *Neuroscience*, *54*, 15-36.

Ekblom, J., Jossan, S. S., Bergström, M., Oreland, L., Walum, E. & Aquilonius, S. M. (1993). "Monoamine oxidase-B in astrocytes." *Glia*, *8*, 122-132.

Ekonomou, A., Savva, G. M., Brayne, C., et al., (2015). "Stage-specific changes in neurogenic and glial markers in Alzheimer's disease." *Biol Psychiatry*, *77*, 711-719.

Engel, P. A. (2014). "Does metabolic failure at the synapse cause Alzheimer's disease?" *Med Hypotheses*, *83*, 802-808.

Escamilla-Tilch, M., Filio-Rodriguez, G., Garcia-Rocha, R., et al., (2013). "The interplay between pathogen-associated and danger-associated molecular patterns: an inflammatory code in cancer?" *Immunol Cell Biol*, *91*, 601-610.

Faideau, M., Kim, J., Cormier, K., et al., (2010). "*In vivo* expression of polyglutamine-expanded huntingtin by mouse striatal astrocytes impairs glutamate transport: a correlation with Huntington's disease subjects." *Hum Mol Genet*, *19*, 3053-3067.

Farkas, E. & Luiten, P. G. (2001). "Cerebral microvascular pathology in aging and Alzheimer's disease." *Prog Neurobiol*, *64*, 575-611.

Fu, W. & Jhamandas, J. H. (2014). "Role of astrocytic glycolytic metabolism in Alzheimer's disease pathogenesis." *Biogerontology*, *15*, 579-586.

Gan, L., Vargas, M. R., Johnson, D. A. & Johnson, J. A. (2012). "Astrocyte-specific overexpression of Nrf2 delays motor pathology and synuclein aggregation throughout the CNS in the alpha-synuclein mutant (A53T) mouse model." *J Neurosci*, *32*, 17775-17787.

Gardner, T. W., Lieth, E., Khin, S. A., et al., (1997). "Astrocytes increase barrier properties and ZO-1 expression in retinal vascular endothelial cells." *Invest Ophthalmol Vis Sci*, *38*, 2423-2427.

Gates, P. E., Strain, W. D. & Shore, A. C. (2009). "Human endothelial function and microvascular ageing." *Exp Physiol*, *94*, 311-316.

Goritz, C., Mauch, D. H. & Pfrieger, F. W. (2005). "Multiple mechanisms mediate cholesterol-induced synaptogenesis in a CNS neuron." *Mol Cell Neurosci*, *29*, 190-201.

Gu, X. L., Long, C. X., Sun, L., Xie, C., Lin, X. & Cai, H. (2010). "Astrocytic expression of Parkinson's disease-related A53T alpha-synuclein causes neurodegeneration in mice." *Mol Brain*, *3*, 12.

Gunčová, I., Látr, I. & Mazurová, Y. (2011). "The neurodegenerative process in a neurotoxic rat model and in patients with Huntington's disease: histopathological parallels and differences." *Acta Histochem*, *113*, 783-792.

Hamm, S., Dehouck, B., Kraus, J., et al., (2004). "Astrocyte mediated modulation of blood-brain barrier permeability does not correlate with a loss of tight junction proteins from the cellular contacts." *Cell Tissue Res*, *315*, 157-166.

Hammond, T. R., Gadea, A., Dupree, J. et al., (2014). "Astrocyte-derived endothelin-1 inhibits remyelination through notch activation." *Neuron*, *81*, 588-602.

Hashioka, S., McLarnon, J. G., Ryu, J. K., et al., (2011). "Pyrazole compound 2-MBAPA as a novel inhibitor of microglial activation and neurotoxicity *in vitro* and *in vivo*." *J Alzheimers Dis*, *27*, 531-541.

Haynes, S. E., Hollopeter, G., Yang, G., et al., (2006). "The P2Y12 receptor regulates microglial activation by extracellular nucleotides." *Nat Neurosci*, *9*, 1512-1519.

Heneka, M. T., Carson, M. J., El Khoury, J., et al., (2015). "Neuroinflammation in Alzheimer's disease." *Lancet Neurol*, *14*, 388-405.

Hogins, J., Crawford, D. C., Jiang, X. & Mennerick, S. (2011). "Presynaptic silencing is an endogenous neuroprotectant during excitotoxic insults." *Neurobiol Dis*, *43*, 516-525.

Holash, J. A., Noden, D. M. & Stewart, P. A. (1993). "Re-evaluating the role of astrocytes in blood-brain barrier induction." *Dev Dyn*, *197*, 14-25.

Hooten, K. G., Beers, D. R., Zhao, W. & Appel, S. H. (2015). "Protective and Toxic Neuroinflammation in Amyotrophic Lateral Sclerosis." *Neurotherapeutics*, *12*, 364-375.

Hsiao, H. Y., Chen, Y. C., Chen, H. M., Tu, P. H. & Chern, Y. (2013). "A critical role of astrocyte-mediated nuclear factor-κB-dependent inflammation in Huntington's disease." *Hum Mol Genetics*, *22*, 1826-1842.

Hughes, E. G., Elmariah, S. B. & Balice-Gordon, R. J. (2010). "Astrocyte secreted proteins selectively increase hippocampal GABAergic axon length, branching, and synaptogenesis." *Mol Cell Neurosci*, *43*, 136-145.

Inglese, M. (2006). "Multiple sclerosis: new insights and trends." *AJNR Am J Neuroradiol*, *27*, 954.

Jebelli, J., Su, W., Hopkins, S., Pocock, J. & Garden, G. A. (2015). "Glia: guardians, gluttons, or guides for the maintenance of neuronal connectivity?" *Ann N Y Acad Sci*, *1351*, 1-10.

Jenner, P. (2003). "Oxidative stress in Parkinson's disease." *Ann Neurol*, *53*, 26-36.

Jessen, K. R. (2004). "Glial cells." *Int J Biochem Cell Biol*, *36*, 1861-1867.

Jha, M. K., Park, D. H., Kook, H., Lee, I. K. & Suk, K. (2015). "Metabolic Control of Glia-Mediated Neuroinflammation." *Curr Alzheimer Res* Epub ahead of print.

Kawamata, H., Ng, S. K., Diaz, N., et al., (2014). "Abnormal intracellular calcium signaling and SNARE-dependent exocytosis contributes to SOD1G93A astrocyte-mediated toxicity in amyotrophic lateral sclerosis." *J Neurosci*, *34*, 2331-2348.

Keaney, J. & Campbell, M. (2015). "The dynamic blood-brain barrier." *FEBS J*, *282*, 4067-4079.

Kettenmann, H. & Verkhratsky, A. (2008). "Neuroglia: the 150 years after." *Trends Neurosci*, *31*, 653-659.

Khakh, B. S. & Sofroniew, M. V. (2014). "Astrocytes and Huntington's disease." *ACS Chem Neurosci*, *5*, 494-496.

Kim, D. K., Kanai, Y., Choi, H. W., et al., (2002). "Characterization of the system L amino acid transporter in T24 human bladder carcinoma cells." *Biochim Biophys Acta*, *1565*, 112-121.

Kim, J., Waldvogel, H. J., Faull, R. L. M., Curtis, M. A. & Nicholson, L. F. B. (2015). "The RAGE receptor and its ligands are highly expressed in astrocytes in a grade-dependant manner in the striatum and subependymal layer in Huntington's disease." *J Neurochem*, *134*, 927-942.

Koehler, R. C., Roman, R. J. & Harder, D. R. (2009). "Astrocytes and the regulation of cerebral blood flow." *Trends Neurosci*, *32*, 160-169.

Krencik, R., Weick, J. P., Liu, Y., Zhang, Z. J. & Zhang, S. C. (2011). "Specification of transplantable astroglial subtypes from human pluripotent stem cells." *Nat Biotechnol*, *29*, 528-534.

Kuchler-Bopp, S., Delaunoy, J. P., Artault, J. C., Zaepfel, M. & Dietrich, J. B. (1999). "Astrocytes induce several blood-brain barrier properties in non-neural endothelial cells." *Neuroreport*, *10*, 1347-1353.

Kucukdereli, H., Allen, N. J., Lee, A. T., et al., (2011). “Control of excitatory CNS synaptogenesis by astrocyte-secreted proteins Hevin and SPARC.” *Proc Natl Acad Sci U S A*, *108*, E440-449.

L'Episcopo, F., Tirolo, C., Caniglia, S., et al., (2014). “Targeting Wnt signaling at the neuroimmune interface for dopaminergic neuroprotection/repair in Parkinson’s disease.” *J Mol Cell Biol*, *6*, 13-26.

L'Episcopo, F., Tirolo, C., Testa, N., et al., (2014). “Wnt/β-Catenin Signaling Is Required to Rescue Midbrain Dopaminergic Progenitors and Promote Neurorepair in Ageing Mouse Model of Parkinson’s Disease.” *Stem Cells*, *32*, 2147-2163.

Lazzarini, M., Martin, S., Mitkovski, M., Vozari, R. R., Stühmer, W. & Bel, E. D. (2013). “Doxycycline restrains glia and confers neuroprotection in a 6-OHDA Parkinson model.” *Glia*, *61*, 1084-1100.

Lee, K. M. & MacLean, A. G. (2015). “New advances on glial activation in health and disease.” *World J Virol*, *4*, 42-55.

Li, P., Yin, Y. L., Li, D., Kim, S. W. & Wu, G. (2007). “Amino acids and immune function.” *Br J Nutr*, *98*, 237-252.

Lian, H., Yang, L., Cole, A., et al., (2015). “NFκB-activated astroglial release of complement C3 compromises neuronal morphology and function associated with Alzheimer’s disease.” *Neuron*, *85*, 101-115.

Lian, H. & Zheng, H. (2016). “Signaling pathways regulating neuron-glia interaction and their implications in Alzheimer’s disease.” *J Neurochem*, *136*, 475-491.

Liu, D., Guo, H., Griffin, J. H., Fernandez, J. A. & Zlokovic, B. V. (2003). “Protein S confers neuronal protection during ischemic/hypoxic injury in mice.” *Circulation*, *107*, 1791-1796.

Liu, Y. & Zhou, J. (2013). “Oligodendrocytes in neurodegenerative diseases.” *Front Biol*, *8*, 127-133.

Lopes Pinheiro, M. A., Kooij, G., Mizee, M. R., et al., (2016). “Immune cell trafficking across the barriers of the central nervous system in multiple sclerosis and stroke.” *Biochim Biophys Acta*, *1862*, 461-471.

Lull, M. E. & Block, M. L. (2010). “Microglial activation and chronic neurodegeneration.” *Neurotherapeutics*, *7*, 354-365.

Mack, A. F. & Wolburg, H. (2013). “A novel look at astrocytes: aquaporins, ionic homeostasis, and the role of the microenvironment for regeneration in the CNS.” *Neuroscientist*, *19*, 195-207.

Madry, C. & Attwell, D. (2015). “Receptors, ion channels, and signaling mechanisms underlying microglial dynamics.” *J Biol Chem*, *290*, 12443-12450.

Malaspina, A., Puentes, F. & Amor, S. (2015). "Disease origin and progression in amyotrophic lateral sclerosis: an immunology perspective." *Int Immunol*, *27*, 117-129.

Markoullis, K., Sargiannidou, I., Schiza, N., Roncaroli, F., Reynolds, R. & A. Kleopa, K. (2014). "Oligodendrocyte gap junction loss and disconnection from reactive astrocytes in multiple sclerosis gray matter." *J Neuropathol Exp Neurol*, *73*, 865-879.

McCarthy, R. C. & Kosman, D. J. (2015). "Mechanisms and regulation of iron trafficking across the capillary endothelial cells of the blood-brain barrier." *Front Mol Neurosci*, *8*, 31.

Meyer, K. & Kaspar, B. K. (2016). "Glia-neuron interactions in neurological diseases: Testing non-cell autonomy in a dish." *Brain Res* Epub ahead of print. doi: 10.1016/j.brainres.2015.12.051.

Miller, G. (2005). "Neuroscience. The dark side of glia." *Science*, *308*, 778-781.

Miyazaki, I., Asanuma, M., Murakami, S., et al., (2013). "Targeting 5-HT(1A) receptors in astrocytes to protect dopaminergic neurons in Parkinsonian models." *Neurobiol Dis 59*, 244-256.

Monteiro-Cardoso, V. F., Oliveira, M. M., Melo, T., et al., (2015). "Cardiolipin profile changes are associated to the early synaptic mitochondrial dysfunction in Alzheimer's disease." *J Alzheimers Dis*, *43*, 1375-1392.

Moore, C. S., Abdullah, S. L., Brown, A., Arulpragasam, A. & Crocker, S. J. (2011). "How factors secreted from astrocytes impact myelin repair." *J Neurosci Res*, *89*, 13-21.

Morgello, S., Uson, R. R., Schwartz, E. J. & Haber, R. S. (1995). "The human blood-brain barrier glucose transporter (GLUT1) is a glucose transporter of gray matter astrocytes." *Glia*, *14*, 43-54.

Mouser, P. E., Head, E., Ha, K. H. & Rohn, T. T. (2006). "Caspase-mediated cleavage of glial fibrillary acidic protein within degenerating astrocytes of the Alzheimer's disease brain." *Am J Pathol*, *168*, 936-946.

Mullett, S. J., Di Maio, R., Greenamyre, J. T. & Hinkle, D. A. (2013). "DJ-1 Expression Modulates Astrocyte-Mediated Protection Against Neuronal Oxidative Stress." *J Mol Neurosci*, *49*, 507-511.

Nagele, R. G., Wegiel, J., Venkataraman, V., Imaki, H., Wang, K. C. & Wegiel, J. (2004). "Contribution of glial cells to the development of amyloid plaques in Alzheimer's disease." *Neurobiol Aging*, *25*, 663-674.

Nedergaard, M., Ransom, B. & Goldman, S. A. (2003). "New roles for astrocytes: redefining the functional architecture of the brain." *Trends Neurosci*, *26*, 523-530.

Niranjan, R. (2014). "The role of inflammatory and oxidative stress mechanisms in the pathogenesis of Parkinson's disease: focus on astrocytes." *Mol Neurobiol*, *49*, 28-38.

Norden, D. M., Fenn, A. M., Dugan, A. & Godbout, J. P. (2014). "TGFbeta produced by IL-10 redirected astrocytes attenuates microglial activation." *Glia*, *62*, 881-895.

Nutt, J. G. & Wooten, G. F. (2005). "Clinical practice. Diagnosis and initial management of Parkinson's disease." *N Engl J Med*, *353*, 1021-1027.

Olmos, G. & Lladó, J. (2014). "Tumor necrosis factor alpha: a link between neuroinflammation and excitotoxicity." *Mediators Inflamm*, 2014, 861231.

Omidi, Y., Barar, J., Ahmadian, S., Heidari, H. R. & Gumbleton, M. (2008). "Characterization and astrocytic modulation of system L transporters in brain microvasculature endothelial cells." *Cell Biochem Funct*, *26*, 381-391.

Osborn, L. M., Kamphuis, W., Wadman, W. J. & Hol, E. M. (2016). "Astrogliosis: An integral player in the pathogenesis of Alzheimer's disease." *Prog Neurobiol* Epub ahead of print. doi: 10.1016/ j.pneurobio. 2016.01.001.

Panaro, M. A. & Cianciulli, A. (2012). "Current opinions and perspectives on the role of immune system in the pathogenesis of Parkinson's disease." *Curr Pharm Des*, *18*, 200-208.

Park, L., Anrather, J., Girouard, H., Zhou, P. & Iadecola, C. (2007). "Nox2-derived reactive oxygen species mediate neurovascular dysregulation in the aging mouse brain." *J Cereb Blood Flow Metab*, *27*, 1908-1918.

Pascual, O., Ben Achour, S., Rostaing, P., Triller, A. & Bessis, A. (2012). "Microglia activation triggers astrocyte-mediated modulation of excitatory neurotransmission." *Proc Natl Acad Sci U S A*, *109*, 197-205.

Paul, D., Ge, S., Lemire, Y., et al., (2014). "Cell-selective knockout and 3D confocal image analysis reveals separate roles for astrocyte-and endothelial-derived CCL2 in neuroinflammation." *J Neuroinflam*, *11*, 10.

Pehar, M., Beeson, G., Beeson, C. C., Johnson, J. A. & Vargas, M. R. (2014). "Mitochondria-Targeted Catalase Reverts the Neurotoxicity of hSOD1G93A Astrocytes without Extending the Survival of ALS-Linked Mutant hSOD1 Mice." *PLoS One*, *9*, e103438.

Phani, S., Re, D. B. & Przedborski, S. (2012). "The Role of the Innate Immune System in ALS." *Front Pharmacol*, *3*, 150.

Phillips, E. C., Croft, C. L., Kurbatskaya, K., et al., (2014). "Astrocytes and neuroinflammation in Alzheimer's disease." *Biochem Soc Trans*, *42*, 1321-1325.

Polazzi, E. & Monti, B. (2010). "Microglia and neuroprotection: from *in vitro* studies to therapeutic applications." *Prog Neurobiol*, *92*, 293-315.

Proschel, C., Stripay, J. L., Shih, C. H., Munger, J. C. & Noble, M. D. (2014). "Delayed transplantation of precursor cell-derived astrocytes provides multiple benefits in a rat model of Parkinsons." *EMBO Mol Med*, *6*, 504-518.

Rojas, F., Gonzalez, D., Cortes, N., et al., (2015). "Reactive oxygen species trigger motoneuron death in non-cell-autonomous models of ALS through activation of c-Abl signaling." *Front Cell Neurosci*, *9*, 203.

Samuraki, M., Matsunari, I., Yoshita, M., et al., (2015). "Cerebral Amyloid Angiopathy-Related Microbleeds Correlate with Glucose Metabolism and Brain Volume in Alzheimer's Disease." *J Alzheimers Dis*, *48*, 517-528.

Savarin, C., Hinton, D. R., Valentin-Torres, A., et al., (2015). "Astrocyte response to IFN-γ limits IL-6-mediated microglia activation and progressive autoimmune encephalomyelitis." *J Neuroinflam 12*, 79.

Schiera, G., Bono, E., Raffa, M. P., et al., (2003). "Synergistic effects of neurons and astrocytes on the differentiation of brain capillary endothelial cells in culture." *J Cell Mol Med*, 7, 165-170.

Sherwood, C. C., Stimpson, C. D., Raghanti, M. A., et al., (2006). "Evolution of increased glia-neuron ratios in the human frontal cortex." *Proc Natl Acad Sci U S A*, *103*, 13606-13611.

Shinozaki, Y., Nomura, M., Iwatsuki, K., Moriyama, Y., Gachet, C. & Koizumi, S. (2014). "Microglia trigger astrocyte-mediated neuroprotection via purinergic gliotransmission." *Sci Rep*, *4*, 4329.

Siddiqui, A., Mallajosyula, J. K., Rane, A. & Andersen, J. K. (2011). "Ability to delay neuropathological events associated with astrocytic MAO-B increase in a Parkinsonian mouse model: Implications for early intervention on disease progression." *Neurobiol Dis*, *43*, 527-532.

Siegel, G. J. & Agranoff, B. W. (1999). "Basic neurochemistry: Molecular, cellular, and medical aspects." Lippincott Williams & Wilkin, Philadelphia, USA.

Sincich, L. C., Horton, J. C. & Sharpee, T. O. (2009). "Preserving information in neural transmission." *J Neurosci*, *29*, 6207-6216.

Stamatovic, S. M., Shakui, P., Keep, R. F., et al., (2005). "Monocyte chemoattractant protein-1 regulation of blood-brain barrier permeability." *J Cereb Blood Flow Metab*, *25*, 593-606.

Stobart, J. L. & Anderson, C. M. (2013). "Multifunctional role of astrocytes as gatekeepers of neuronal energy supply." *Front Cell Neurosci*, *7*, 38.

Tasdemir-Yilmaz, O. E. & Freeman, M. R. (2014). "Astrocytes engage unique molecular programs to engulf pruned neuronal debris from distinct subsets of neurons." *Genes Dev*, *28*, 20-33.

Thangavelu, S. R., Tripathi, P. P., Arya, U., Mishra, H. K. & Subramaniam, J. R. (2011). "ALS associated mutant SOD1 impairs the motor neurons and astrocytes and wild type astrocyte secreted-factors reverse the impaired motor neurons." *Ann Neurosci*, *18*, 48-55.

Thomsen, M. S., Andersen, M. V., Christoffersen, P. R., Jensen, M. D., Lichota, J. & Moos, T. (2015). "Neurodegeneration with inflammation is accompanied by accumulation of iron and ferritin in microglia and neurons." *Neurobiol Dis*, *81*, 108-118.

Thouin, A. & Crompton, D. E. (2016). "Glut1 deficiency syndrome: Absence epilepsy and La Soupe du Jour." *Pract Neurol*, *16*, 50-52.

Tong, X., Ao, Y., Faas, G. C., et al., (2014). "Astrocyte Kir4.1 ion channel deficits contribute to neuronal dysfunction in Huntington's disease model mice." *Nat Neurosci*, *17*, 694-703.

Torkildsen, Ø., Brunborg, L. A., Myhr, K. M. & Bø, L. (2008). "The cuprizone model for demyelination." *Acta Neurol Scand Suppl*, 117, 72-76.

Tortarolo, M., Vallarola, A., Lidonnici, D., et al., (2015). "Lack of TNF-alpha receptor type 2 protects motor neurons in a cellular model of amyotrophic lateral sclerosis and in mutant SOD1 mice but does not affect disease progression." *J Neurochem*, *135*, 109-124.

Tovar-Y-Romo, L. B., Ramírez-Jarquín, U. N., Lazo-Gómez, R. & Tapia, R. (2014). "Trophic factors as modulators of motor neuron physiology and survival: implications for ALS therapy." *Front Cell Neurosci*, *8*, 61.

Ueno, M., Chiba, Y., Matsumoto, K., et al., (2015). "Blood-brain barrier damage in vascular dementia." *Neuropathology* Epub ahead of print. doi: 10.1111/neup.12262.

Ullian, E. M., Sapperstein, S. K., Christopherson, K. S. & Barres, B. A. (2001). "Control of synapse number by glia." *Science*, *291*, 657-661.

Valenza, M., Marullo, M., Di Paolo, E., et al., (2015). "Disruption of astrocyte-neuron cholesterol cross talk affects neuronal function in Huntington's disease." *Cell Death Differ*, *22*, 690-702.

Vana, A. C., Flint, N. C., Harwood, N. E., Le, T. Q., Fruttiger, M. & Armstrong, R. C. (2007). "Platelet-derived growth factor promotes repair of chronically demyelinated white matter." *J Neuropath Exp Neurol*, *66*, 975-988.

Venigalla, M., Sonego, S., Gyengesi, E., Sharman, M. J. & Munch, G. (2015). "Novel promising therapeutics against chronic neuroinflammation and neurodegeneration in Alzheimer's disease." *Neurochem Int* Epub ahead of print. doi: 10.1016/j.neuint.2015.10.011.

Verkhratsky, A., Parpura, V., Pekna, M., et al., (2014). "Glia in the pathogenesis of neurodegenerative diseases." *Biochem Soc Trans*, *42*, 1291-1301.

Verkman, A. S. (2012). "Aquaporins in clinical medicine." *Annu Rev Med*, *63*, 303-316.

Wallach, G., Lallouette, J., Herzog, N., et al., (2014). "Glutamate mediated astrocytic filtering of neuronal activity." *PLoS Comput Biol*, *10*, e1003964.

Wang, Q., Liu, Y. & Zhou, J. (2015). "Neuroinflammation in Parkinson's disease and its potential as therapeutic target." *Transl Neurodegener*, *4*, 19.

Willis, C. L., Leach, L., Clarke, G. J., Nolan, C. C. & Ray, D. E. (2004). "Reversible disruption of tight junction complexes in the rat blood-brain barrier, following transitory focal astrocyte loss." *Glia*, *48*, 1-13.

Wójtowicz, A. M., Dvorzhak, A., Semtner, M. & Grantyn, R. (2013). "Reduced tonic inhibition in striatal output neurons from Huntington mice due to loss of astrocytic GABA release through GAT-3." *Front Neural Circuits*, *7*, 188.

Won, S. J., Yoo, B. H., Kauppinen, T. M., et al., (2012). "Recurrent/moderate hypoglycemia induces hippocampal dendritic injury, microglial activation, and cognitive impairment in diabetic rats." *J Neuroinflam*, *9*, 182.

Woodman, S. E., Benveniste, E. N., Nath, A. & Berman, J. W. (1999). "Human immunodeficiency virus type 1 TAT protein induces adhesion molecule expression in astrocytes." *J Neurovirol*, *5*, 678-684.

Xu, J., Xiao, N. & Xia, J. (2010). "Thrombospondin 1 accelerates synaptogenesis in hippocampal neurons through neuroligin 1." *Nat Neurosci*, *13*, 22-24.

Yamamoto, M., Kamatsuka, Y., Ohishi, A., Nishida, K. & Nagasawa, K. (2013). "P2X7 receptors regulate engulfing activity of non-stimulated resting astrocytes." *Biochem Biophys Res Commun*, *439*, 90-95.

Yanagida, O., Kanai, Y., Chairoungdua, A., et al., (2001). "Human L-type amino acid transporter 1 (LAT1): characterization of function and expression in tumor cell lines." *Biochim Biophys Acta*, *1514*, 291-302.

Yates, D. (2015). "Neurodegenerative disease: Factoring in astrocytes." *Nat Rev Neurosci*, *16*, 67-67.

Youssef, A. M., White, M. S., Villanueva, E. B., El-Ashmawy, I. M. & Klegeris, A. (2010). "Synthesis and biological evaluation of novel pyrazolyl-2,4-thiazolidinediones as anti-inflammatory and neuroprotective agents." *Bioorg Med Chem*, *18*, 2019-2028.

Zhao, Z., Nelson, A. R., Betsholtz, C. & Zlokovic, B. V. (2015). "Establishment and Dysfunction of the Blood-Brain Barrier." *Cell*, *163*, 1064-1078.

Zhu, D., Wang, Y., Singh, I., et al., (2010). "Protein S controls hypoxic/ischemic blood-brain barrier disruption through the TAM receptor Tyro3 and sphingosine 1-phosphate receptor." *Blood*, *115*, 4963-4972.

In: New Developments in Astrocytes Research ISBN: 978-1-63485-561-7
Editor: Leyla Estrada

Chapter 3

ASTROCYTES IN AMYOTROPHIC LATERAL SCLEROSIS: HARMFUL EFFECTS AND STRATEGIES TO AFFORD NEUROPROTECTION

***A. R. Vaz*[1,2]*, C. Gomes*[2]*, C. Cunha*[2] *and D. Brites*[1,2]**

[1]Department of Biochemistry and Human Biology and
[2]Research Institute for Medicines, Faculty of Pharmacy,
University of Lisbon, Lisbon, Portugal

ABSTRACT

Amyotrophic lateral sclerosis (ALS) is the third most common neurodegenerative disease, rapidly progressing and fatal. ALS is characterized by the loss of motor neurons, although neuroinflammation and glial cell activation are also part of the pathological underlying mechanisms. Indeed, phenotypically dysfunctional astrocytes with toxic properties towards motor neurons were identified in humans with either sporadic or familial ALS, as well as in transgenic rodent models, but the mechanisms involved are far from being completely clarified. Since no adequate treatment is available, intraspinal transplantation of neural stem cells to prevent motor neuron loss in ALS has been attempted to provide neuroprotection and/or replace lost cells. However, therapeutic benefits were scarce and limited to particular time frames when translated into clinical trials, which may derive from the neurotoxic properties of local astrocytes secreting unknown neurotoxins at the time of transplantation. Here, we summarize the current understanding on the role of astrocytes in ALS pathology, highlighting their deregulated protein signaling pathways, the associated neuroinflammatory biomarkers, as well as the gain of toxic functions along disease progression. In addition, we point out the contribution that soluble factors, extracellular microvesicles (ectosomes) and exosomes, as part of the astrocyte secretome, may have in disturbing normal cell-to-cell communication. Such circulating vesicles due to their content in specific proteins, mRNAs and microRNAs, have been associated to the dissemination of misfolded, mutant proteins, neurotoxins and neuroinflammatory mediators in neurodegenerative diseases, such as ALS. Finally, we address the potential benefits of astrocyte-targeted approaches for motor neuron protection in ALS, focusing on replacement-based therapies that may use astrocytes generated from induced pluripotent stem cells (iPSCs) or transdifferentiated from fibroblasts collected in patients with ALS.

1. INTRODUCTION

Amyotrophic Lateral Sclerosis (ALS) is the most common adult-onset neurodegenerative disease affecting motor neurons (MNs). It is characterized by the selective loss of MNs in the motor cortex (upper MNs), as well as in the brainstem and in the spinal cord (lower MNs). The first cases of MN degeneration were described between 1865 and 1869 by the neurobiologist and clinician Jean-Martin Charcot. It is now known that the symptoms vary depending on the location of the lesion [1]. ALS onset is within the upper limb, lower limb or bulbar musculature, with a high variable disease progression rate [2]. Bulbar symptoms that usually progresses faster than limb onset, include as first signs the existence of dysphagia, dysarthria and a slurred speech that are observed in up to 30% of ALS patients [3], with upper and lower limb-onset equally represented [2]. Most people with ALS show muscle weakness, spasticity, atrophy and ultimately, paralysis, culminating in death typically by respiratory failure contractures [4]. Usually, the cognitive functions of the brain remain undamaged and the patient is aware of the disease progression. The disorder commonly has a fast rate of progression being the typical age of onset around 48 years (although it can occur early in life) and patients usually die within 3-5 years from symptoms onset. ALS has a worldwide incidence of 1 to 2 new cases per 100,000 individuals each year and a prevalence of 4 to 6 cases per 100,000 individuals [5]. Concerning etiology, about 90% of the ALS cases have an unknown cause and are non-genetic, thereby classified as sporadic ALS (sALS). The remaining 5 to 10% of cases are from genetic origin, known as familial ALS (fALS) due to several genetic mutations. A recurrent feature with other neurodegenerative diseases is the formation of protein aggregates in the degenerating MNs. Among the most common genes identified so far, are those encoding for superoxide dismutase 1 (SOD1), transactive response DNA-binding protein (TARDBP), fused in sarcoma (FUS) and C9ORF72. Mutations in SOD1 may originate distal axonopathy, being neuromuscular junctions the first regions of the MNs to degenerate in a process known as 'dying back' [6]. It is believed that a gain of toxic function mechanism is the main pathway for SOD1 neurotoxicity [7]. The acquisition of the toxic function might reside in the propensity of mutant enzymes to misfold and aggregate caused by decreased stability of the native homodimer or a tendency to lose stabilizing posttranslational modifications [8, 9]. Lately, cytotoxicity associated with SOD1 aggregation in ALS was suggested to be produced by a non-native trimeric SOD1 species, providing a direct link between the presence of misfolded oligomers and neuron death [10]. However, the relevance of such finding in the context of the molecular mechanisms by which SOD1 mutations trigger the disease are still to be discovered. While mice lacking SOD1 did not develop the disease [11], transgenic mice overexpressing human mutant SOD1 developed ALS clinical features [12]. Several studies indicate that both genetic and non-genetic forms of ALS triggering a compromised homeostasis of MNs share key pathogenic mechanisms such as protein aggregation [13], endoplasmic reticulum stress, mitochondrial failure and axonal transport dysfunction [14, 15]. Indeed, as depicted in Table 1, several mutations were identified in both fALS and sALS cases. A main hallmark of ALS is MN degeneration and identification of SOD1 mutations has enabled development of animal and cell culture models from which it was possible to uncover some of the mechanisms triggering MN injury, i.e., oxidative stress, glutamate excitotoxicity, protein aggregation, endoplasmic reticulum stress, mitochondrial dysfunction, axonal transport abnormalities, and neuroinflammation [13-16]. Actually, deffects in axonal transport were shown to be a common factor contributing to the ALS pathogenic process [17]. The increase of the phosphoneurofilament heavy chain and the neurofilament light chain in cerebrospinal fluid and in blood are now considered emerging molecular biomarker targets to measure disease progression [18-21]. Nevertheless, their application in the disease first stages does not allow differentiation between ALS and non-ALS patients. Other studies have pointed to the loss of major histocompatibility complex class I

(MHCI) molecules on MNs as triggering MN susceptibility to astrocyte-induced cell death [22]. Moreover, death of MNs by astrocytes from sALS and fALS samples, using humanized cell coculture models, revealed the necroptotic pathway as also driving the MN death [23].

Table 1. ALS-associated mutations in both familial ALS (fALS) and sporadic ALS (sALS)

Mutation	Inheritance	% of fALS	First reported in fALS	Reported in sALS
Superoxide dismutase 1 (SOD1)	Autossomal Dominant	20%	[176]	[15, 177]
TAR DNA-binding protein 43 (TDP-43)	Autossomal Dominant	4%	[178]	[179, 180]
Fused in sarcoma (FUS)	Autossomal Dominant	1-5%	[181]	[182]
Angiogenin (ANG)	Autossomal Dominant	<1%	[183]	[183]
Optineurin (OPTN)	Autossomal Dominant	<1%	[184]	[185]
Chromosome 9 open reading frame 72 (C9ORF72)	Autossomal Dominant	20-50%	[186]	[187]

The generation of MNs from induced pluripotent stem cells (iPSCs) or neuronal progenitor cells (NPCs) derived from the skin fibroblasts of ALS patients is expected to be a major advance in modelling pathogenesis in ALS, with potential for high-throughput therapeutic assessments. iPSCs from patients with fALS have already been made available to researchers at the NINDS Repository, thus allowing for the study of ALS arising from SOD1, FUS, and FIG4 mutations. It is now widely accepted that iPSCs can reproduce features of MN vulnerability in ALS, including protein aggregation and reduced cell survival [24], and NPC-derived astrocytes were shown to induce toxicity towards MNs [25]. Moreover, the generation of iPSCs from diverse patients, either sALS or fALS, is expected to engender novel criteria for risk and patient stratification, as well as drug testing.

Given the variety of mechanisms reported to be involved in the disease, it seems that ALS results from cumulative upstream abnormalities which culminate in a common pathogenic cascade. Besides MN injury, it is now accepted that the surrounding glial cells are also actively implicated in the disorder. By using transgenic animals with cell type specific manipulations of mutant SOD1 expression, it was possible to understand that unlike the traditional SOD1 mouse models [26], which display ubiquitous expression of mutant SOD1, selective expression of mutant SOD1 only in neurons [27], or in astrocytes [28], was not sufficient to cause ALS-like pathology in mice. Inhibition of mutant SOD1 specifically in neurons significantly delayed disease onset, but had little effects on disease progression, while selective inhibition of mutant SOD1 in microglia, or in astrocytes, had no incidence on disease onset, but significantly delayed disease progression [29, 30].

More recently, several studies indicate that astrocytes from both fALS and sALS cases have toxicity towards MNs [23, 25]. It was additionally evidenced that the direct conversion of induced neural precursor cells from fibroblasts of ALS patients with the two most frequent mutations [mutated form (G93A) of Cu, Zn superoxide dismutase (mSOD1) and C9ORF72] were as toxic as those from sALS cases [25], pointing to a common mechanism. This may be due to the propagation and transmission of misfolded SOD1, now considered the cause of most, if not all, types of ALS [31]. In light of the central role played by astrocytes in the function of the central nervous system (CNS), it is not surprising that they have been implicated either in the onset or in the progression of ALS. The question of whether the involvement of astrocytes is a consequence of the loss of their supportive roles (loss-of-function), or of a toxic gain-of-function, or both, is currently under active investigation.

Recent research suggests that astrocytes secrete substances that contribute to the regulation of CNS development and homeostasis, synaptogenesis and cognitive function [32]. Different molecules may be released from astrocytes through distinct pathways, such as vesicle-based exocytosis, diffusion via plasmalemmal pores/channels and extrusion mediated by plasmalemmal transporters, as reviewed by Verkhratsky and colleagues [33]. Notoriously, astrocytes may release the same molecule by one or more of these pathways at the same time, which may account for the propagation and transmission of misfolded proteins, such as SOD1 [31]. Astrocytes secrete neurotransmitters, neuromodulators, hormones and growth factors, all of which contributing to the gliocrine system [34]. Evidence is accumulating that released extracellular vesicles (EVs), comprising exosomes and ectosomes, may carry specific proteins, mRNAs and microRNAs (miRNAs) [35, 36]. However, such mediated neurotoxic processes, either exerted by soluble factors [37], or through the astrocyte derived exosomes/ectosomes [31, 38], are only at the beginning of knowledge and needs to be clarified.

No effective treatment is currently available for ALS and patient care focuses mainly on symptomatic treatments and physical therapy. Assisted ventilation and nutrition can transiently overcome the loss of upper airway and respiratory muscular control [4]. In spite of the large number of therapeutic trials, the only drug approved by Food and Drug Administration (FDA) for the treatment of patients with ALS is riluzole, an antiglutamatergic agent that blocks the presynaptic release of glutamate. However, its efficacy is limited, since minimal therapeutic benefits are observed, with only about 3-4 months of survival increase [39]. Consistent with the important role of astrocytes in the pathophysiology of ALS, recent studies in ALS animal models demonstrate that targeting astrocytic functions may be a promising therapeutic avenue. Therefore, it is essential to dissect the molecular mechanisms underlying MN loss in ALS, not only in terms of MN degeneration but also looking for the neurotoxic role of astrocytes through the release of harmful toxins. Indeed, there is a growing body of evidence supporting that astrocytopathies associated with astrodegeneration or dysfunctional astrogliosis are primary causes or major factors in neurological dysfunction and disease [34].

Here, we summarize the current understanding on the role of astrocytes in ALS pathology, highlighting the deregulated signaling pathways in dysfunctional astrocytes, and the toxicity exerted by their secretome, including the production of soluble factors, extracellular microvesicles and exosomes. Finally, we address the potential benefits of astrocyte-targeted approaches for MN protection in ALS, which may open novel avenues for therapeutic intervention.

2. Involvement of Astrocyte Dysfunction Along ALS Progression

Astrocytes play a critical role in the viability and function of the CNS, either in neurodevelopment or in neurodegeneration. It has been widely reviewed that astrocytes have integral roles for the formation, maintenance, and elimination of synapses [40], supplying neurons with essential metabolites and actively participating in the maintenance of important homeostatic functions, such as glutamate buffering and tissue repair after injury [41, 42]. Therefore, it is not surprising that the deregulation of these multifunctional cells may be implicated in the onset and progression of several neurodegenerative diseases, such as ALS. The contribution of astrocytes for the neurotoxic environment that is present in ALS is likely a consequence of both the loss of normal homeostatic functions and gain of toxic functions of these cells.

2.1. Molecular Biomarkers in ALS

Many efforts have been conducted to develop reliable biomarkers to early diagnose ALS mainly focused on either biochemical or live-imaging fields [43]. However, due to the absence of specific disease markers, clinical diagnostic criteria has been mainly based on the presence and distribution of lower and upper MN failure and quantification of recognized ALS clinical phenotypes in terms of respiratory and electrophysiological parameters. The diagnosis of ALS currently depends on the opinion of an experienced neurologist, and the exclusion of other potential MN diseases. However, these clinical and functional measures alone may not be enough as indicators of the biological activity of the disease.

The contribution of genetic factors to the clinical phenotype and rate of progression in ALS is incompletely understood and, in spite of no obvious relationship between underlying genetic cause and phenotype, both genetic and non-genetic forms of ALS share common pathogenic mechanisms [44]. The striking functional and structural similarities of TDP-43 and FUS, both DNA/RNA binding proteins, suggest that abnormal RNA processing and protein aggregation are crucial events in ALS pathogenesis. Although to date most cases are unrelated to mutations in SOD1 gene, the chance that sALS and fALS may converge into a common pathogenic pathway involving abnormal SOD1 species must be considered. For instance, SOD1 protein aggregates were found in the spinal cord of human fALS and sALS cases [13], and wild-type SOD1 (wtSOD1) was shown to acquire binding and toxic properties of mSOD1 through oxidative damage [14, 45]. Also, wtSOD1 can interact with Heat shock protein 70 (Hsp70), as well as with chromogranin B, inducing tumor necrosis factor-α (TNF-α) and inducible nitric oxide synthase (iNOS), thereby leading to a dose dependent cell death [14]. Moreover, oxidized wtSOD1 and mSOD1 were reported to share a conformational epitope that is not present in normal wtSOD1, and both recombinant oxidized wtSOD1 and mSOD1 immunopurified from human sALS tissues have been shown to inhibit kinesin-based fast axonal transport, similarly to SOD1-fALS [15]. Cellular markers of neuroinflammation in ALS, as a result of activated astrocytes and microglia, followed by infiltrated T-lymphocytes, are considered one of the main ALS features, as evidenced in human postmortem samples and in animal models of the disease (reviewed in [46, 47]). These findings suggest that neuroinflammation in ALS is not a transient feature, but rather a permanent issue. The immune response, although having beneficial effects at early stages of disease progression, go into a vicious cycle contributing to additional neuronal damage by the sustained presence of pro-inflammatory mediators. Several studies have demonstrated elevated circulating immune complexes (CICs) in ALS patients. Lately, Saleh and coworkers [48] suggested a humoral immune response over disease progression, as they found elevated CICs and IgG in serum of ALS patients at the initial disease stages. Intriguingly, in ALS patients with lower functional rating score, the Authors found that levels of CICs were inferior to controls, whereas IgG remained elevated.

The increased levels of other pro-inflammatory markers are also described in spinal cord and cerebrospinal fluid of ALS patients. In this respect, both the regulatory calcium (Ca^{2+}) binding protein S100B that modulates the activity of effector proteins and cells, and the soluble form of CD14 released by monocytes and involved in inflammation, are associated with glial reactivity [49-51]. Other inflammatory-associated markers have been proposed in the context of ALS, such as: (i) the S100A6 overexpression in human patients, mouse models and within astrocytes [52]; the increase of matrix metalloproteinase-9 (MMP-9) in MNs [53], animal models [53, 54], and in serum of ALS patients [55]; and decreased creatinine levels together with increased ferritin levels in sALS patients [56]. The study of Süssmuth and colleagues [50] revealed that only a minority of MNs presented S100B staining, while most of the S100B-positive cells were reactive astrocytes, indicating that reactive astrogliosis occurs along disease progression. In addition, they reported that cerebrospinal fluid concentrations of S100B were inversely correlated with survival time in ALS patients. Other works have shown that advanced glycation end-products (AGEs) levels were higher in mSOD1 than in control human or in mouse samples [57]. Nevertheless, the S100B receptor RAGE was found significantly decreased in the serum of ALS patients [58], suggesting that low soluble RAGE levels are a risk factor for ALS.

Astrogliosis was shown to also occur in the course of maladaptative changes in adenosine homeostasis by overexpression of adenosine kinase in astrocytes [59]. Recently the gap junction protein connexion 43 (Cx43) in astrocytes was shown to contribute to MN toxicity in ALS [60]. Increased Cx43 was found in the cortex and spinal cord of ALS patients and in astrocytes isolated from the mSOD1 mice model and generated from IPSCs. Another phenomenon that may implicate astrocyte loss-of-function in the context of ALS is excitotoxicity. Excitotoxicity is a well-known mechanism implicated in neuronal damage resulting from the excessive influx of Ca^{2+} through glutamate receptors such as N-methyl-D-aspartate (NMDA). The event was proposed many years ago due to observations of increased levels of glutamate in the serum and cerebrospinal fluid of ALS patients [61]. In addition, Texido and colleagues [62] reported that sera from ALS patients induce abnormal NMDA receptor activation. However, excitotoxicity may also result from alterations in glutamate clearance at the synaptic cleft. Excitatory amino-acid transporters (EAATs) are the main regulators of extracellular glutamate concentrations in the CNS and the glial carriers EAAT1 and EAAT2 have the highest impact in removing glutamate from the synaptic cleft [63]. Deficits in these glutamate transporters have been identified in the motor cortex and the spinal cord of ALS patients, especially the astroglial glutamate transporter EAAT2 [64]. Loss of EAAT2 was also reported in mSOD1 animal models of ALS [65] and in astrocyte cultures from the spinal cord of symptomatic mSOD1 rats [37]. Reduced expression was similarly observed in the brain cortex of mSOD1 mice at the symptomatic stage and in astrocytes isolated from their pups (our unpublished data). This issue will be further explored in the next section in the context of the inflammatory mediators associated with reactive/dysfunctional astrocytes.

2.2. Inflammatory Mediators and Biomarkers in the SOD1 Mice Model of ALS

Reactive astrogliosis in ALS was associated with increased glial fibrillary acidic protein (GFAP) and S100B expression, as well as with elevated levels of the inflammatory marker cyclo-oxygenase-2 (COX-2) (reviewed in [66]). However, there is some literature controversy related with increased [30], or decreased [37] GFAP levels in ALS experimental models and novel information has been added to these findings based on recent studies. Disruption of glutamate uptake, in either proliferating astrocytes of the spinal cord and motor cortex of mSOD1 rodents or in ALS patients, has been considered a hallmark of ALS and a cause for ROS production by MNs [67, 68]. However, focal restoration of EAAT2 in the mSOD1 mice did not delay forelimb disease onset, and was unable to prolong overall animal survival [69]. Due to the decreased ability of mSOD1 astrocytes to regulate the neuronal glutamate receptor (GluR2) [a subunit of α-amino-3-hydroxy-5-methyl-4-isoxazole propionic acid (AMPA) receptor in MNs], MNs are submitted to high concentrations of Ca^{2+} and excitotoxicity, with subsequent cell death [70]. Moreover, reactive ALS astrocytes are immunoreactive to nitrotyrosine and express high levels of iNOS and NO, attesting oxidative and nitrosative stress [71, 72]. Mitochondrial dysfunction was also identified in astrocytes isolated from mSOD1 rats and was shown to critically influence MN survival [73].

Haidet-Phillips et al., [74] have reported the increased gene expression of chemokines and pro-inflammatory cytokines in astrocytes of sALS and fALS patients. Accordingly, transforming growth factor-β 1 (TGF-β1) was described to be increased in astrocytes from ALS patients and rodent ALS models. TGF-β1 negatively regulates the potential immune response by microglia and T cells and promotes ALS progression [75]. The mSOD1 astrocytes were shown to express higher levels of TNF-α, interferon gamma (IFNγ), interleukins (IL)-1β and IL-18, 5-lipoxygenase (5-LOX), leukotriene B4, COX-2 and prostaglandin E_2 (PGE_2), when compared to control cells [76, 77]. Recently, the mRNA levels of intracellular NOD-like receptor family pyrin domain containing 3(NLRP3) were found increased in the spinal cord of symptomatic mSOD1 mice [78], namely in astrocytes, a finding also observed by us (unpublished data). NLRP3, as part of the innate immune system, promotes the activation of caspase-1 and the consequent maturation of IL-1β and IL-18 from their proforms [78, 79]. How the astroglial NLRP3 inflammasome complex is involved in neuroinflammation in ALS needs to be clarified.

Another aspect of neuroinflammation and neurodegeneration is the recently observed dysregulation of miRNA expression transversal to different inflammatory and MN disorders [80, 81]. MiRNAs are small (~22 nucleotides), single stranded non-coding RNAs involved in regulation of gene expression at a post-transcriptional level. They directly bind to the 3' untranslated region (UTRs) of a specific mRNA inducing repression of protein synthesis or mRNA degradation. Since miRNA binding to a target mRNA is most of the times incomplete, a single miRNA can target multiple miRNAs and a specific transcript may be regulated by different miRNAs, creating complex networks of gene expression regulation [82]. In the context of ALS, the dysregulation of several miRNAs have been identified, including that of miRNA(miR)-206, miR-9, miR-124a, miR-155, miR-22 and miR-146b, affecting either muscle, MNs or glial cells [81, 83, 84]. Particularly, miRNAs associated with inflammation, often called inflamma-miRs, seem to be among the most upregulated species in ALS patients. Among them, miR-155, which is a potent amplifier of inflammation, was shown to be increased in the spinal cord of ALS patients [85] and in transgenic mice models [86, 87], even before symptom onset (as recently observed by us). Moreover, the observed decrease in miR-146a expression in the cortex of mSOD1mice model may be the causative factor for the enhanced release of the high mobility group box 1 (HMGB1) we found. This alarmin is able to induce inflammatory responses upon Toll-like receptors (TLRs) binding [88].

Notably, a distinct subpopulation of reactive and proliferative astrocytes was identified in rodent ALS models and in patients. Atypical reactive astrocytes were first identified in the spinal cord of a patient with both ALS and glioblastoma [89], and afterwards in the primary motor cortex and gray matter of sALS and fALS cases [90, 91]. Dysfunctional astrocytes with a marked proliferative capacity, deficient replicative senescence and evidencing distinctive antigenic markers of undifferentiated cells, including high expression of S100B and Cx43, together with low levels of GFAP and EAAT2, were observed in astrocyte cultures isolated from the spinal cord of symptomatic mSOD1 rats [37]. This subpopulation of phenotypically aberrant astrocytes was indicated to have 10-fold higher neurotoxic potential, than neonatal mSOD1 astrocytes [37]. More recently, data obtained in our laboratory characterized a subtype of astrocytes isolated from the brain cortex of mSOD1 mice pups has also having neurotoxic effects in the MN-like cell line NSC-34 (unpublished observation). Actually, such neurotoxic phenotype is similar to reactive astrocytes observed *in vivo* [92] and manifests before any reactive gliosis occurs.

Remarkably, the transplantation of rodent glial progenitors in mSOD1 mice just slightly extended animal survival [93, 94], while the use of human glial progenitors did not produce any improvement in disease phenotype [95]. The recurrent failure of these cell-based therapeutic approaches may reside on the presence of local reactive and dysfunctional astrocytes with elevated toxicity and destructive potential.

In conclusion, the involvement of astrocytes in the pathology of this neurodegenerative disease is likely caused by a combined effect resulting from loss of normal homeostatic functions, gain of toxic functions and exacerbated reactivity of such cells in ALS leading to cell demise. Thus, the identification of factors and processes that serve as initiators, modulators, or executioners of MN degeneration in ALS, together with the discovery of drugs effective to target phenotypically aberrant astrocytes, are of major relevance to find suitable biomarkers for ALS early diagnosis and for screening promising neuroprotective drugs for this pathology.

3. Contribution of Astrocytes and Deregulated Cross-Talk for MN Degeneration in ALS

Although the exact mechanism by which astrocytes promote MN death in ALS is unknown, it is recognized that alterations in astrocyte normal function are implicated in disease onset and progression. Actually, cell-type sensitive genetic manipulations of the mouse model of ALS demonstrated that astrocytes and microglia are associated to the rate of progression of fALS with astrocytes determining the timing of microglial activation [29, 30]. In particular, astrocytes were evidenced to be highly reactive (Díaz-Amarilla et al., 2011) and toxic to MNs (Díaz-Amarilla et al., 2011; Meyer et al., 2014; Re et al., 2014). However, it remains to be clarified whether all astrocytes are intrinsically neurotoxic for MNs or if only a specific subtype is implicated. Insights into the complex intercellular perturbations underlying ALS, including target identification and the role of microglia and cell interconnectivity, were reviewed by Brites and Vaz [96]. These Authors described microglia cell phenotypes during neuroinflammation with a focus on the inflammatory miRNA profiling. Astrocytes also undergo diverse phenotypic changes in health and disease [97]. Astrocytes with ability to produce proinflammatory mediators were termed A1 and immunoregulatory mediators as A2 [98]. Importantly, mouse and human astrocytes showed to differently respond to IL-1, confirming that differential innate immune activation pathways are involved and that translation of astrocytes data derived from rodents to humans should be cautiously undertaken. This may be one of the reasons of disparate results obtained by transplantation of stem-cell-derived astrocytes in ALS [99].

In this section it will be detailed the mechanisms by which astrocytes exert direct and selective toxicity to MNs, either by ligand-receptor cross-talk or by secreting soluble factors and/or vesicles. A summary of the most relevant molecules implicated in the astrocyte toxicity towards MNs in ALS is depicted in Figure 1.

3.1. Cell-to-Cell Communication

Several *in vitro* studies demonstrated that mouse [92, 100] or human stem cell-derived [101, 102] MNs carrying human wild-type SOD1 undergo degeneration when cultured with astrocytes carrying SOD1 mutation.

Remarkably, the survival of other types of neurons, including interneurons or dorsal root ganglion neurons, was not affected by the presence of mSOD1 astrocytes. In addition, co-culture of MNs with other cell types expressing mSOD1, namely fibroblasts, microglia, cortical neurons and myocytes, did not induce neurodegeneration [92, 102]. Together, these studies indicate the presence of astrocyte-specific mechanisms contributing to mSOD1-induced MN degeneration.

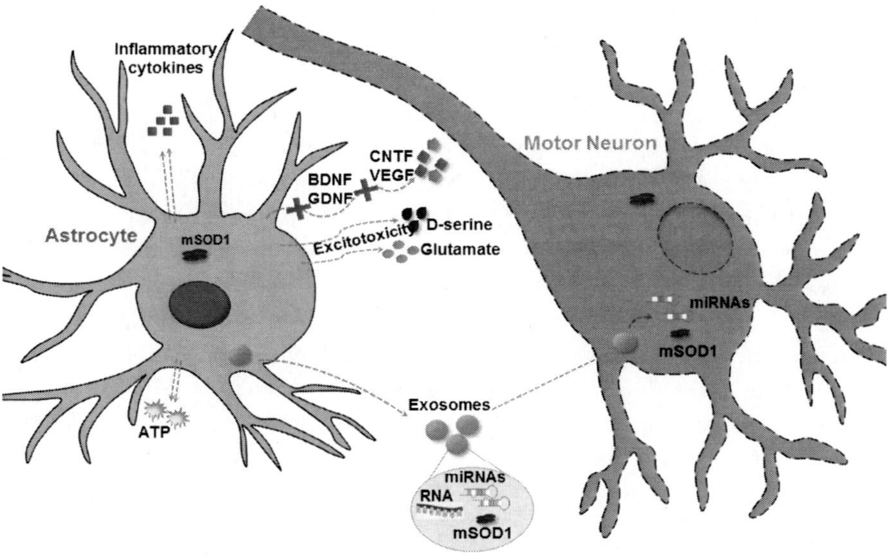

Figure 1. Schematic representation of the mechanisms involved in the motor neuron (MN) degeneration by dysfunctional astrocytes in ALS. Astrocytes lose the capacity of secreting neurotrophic factors, such as brain-derived neurotrophic factor (BDNF), glial-derived neurotrophic factor (GDNF), ciliary neurotrophic factor (CNTF), and vascular endothelial growth factor (VEGF), which are essential for MN survival. On the other hand, the high levels of D-Serine and glutamate secreted by astrocytes into the extracellular environment were shown to contribute to excitotoxicity and consequent MN degeneration. ALS-related astrocytes additionally release high concentrations of several pro-inflammatory cytokines and ATP, thus contributing to the astrocyte-induced neurotoxicity. In such neurodegenerative network, astrocyte-derived exosomes are also important players, since they may transport microRNAs (miRNAs) and mutant SOD1 (mSOD1), which were shown to be internalized by MNs and to be mediators of neurodegeneration.

Although the exact mechanisms by which MN degenerates are not still understood, several non-exclusive mechanisms have been proposed, such as oxidative and nitrosative stress, increased microglial activation and/or excitotoxicity, as reviewed by Allaman and colleagues [41]. In addition, alterations in astrocytes may include elevated levels of GFAP-positive astrocytes, which were found in mSOD1 mice models prior to disease onset, and later on increased prominence during disease progression [30]. However, a subpopulation of astrocytes has ben related with immature and reactive astrocytes, presenting low GFAP and high S100B levels[37].

Increased extracellular levels of ATP were found in ALS-related MNs in culture [54] and in astrocyte supernatants after stimuli, which are known to affect neuronal activities, once ATP is implicated in the regulation of synaptic transmission and in dynamic communication between astrocytes and neurons [103]. Indeed, ATP and glutamate were shown to coordinately activate astrocytes, through their internal Ca^{2+} content, leading to the release of such neuroactive molecules, known to be involved in both inhibitory and excitatory effects on neighboring neurons [104].

It was recently proposed that TDP-43 aggregation and misfolding in astrocytes of the spinal cord contribute to MN degeneration [105]. Both mSOD1 and TDP-43 derived astrocytes cause the death of MNs trough a common pathogenic pathway that involves nitrosative stress [106]. Recent findings also highlight that TLR2 and TLR4 are critical for glia activation. While TLR2 was predominantly detected in microglia, TLR4 was strongly expressed in astrocytes in the spinal cord of ALS patients [107]. MN degeneration by TLR2 overexpression may be envisaged based on the inhibition of neural progenitor cell proliferation by TLR2 [108] and on its association with synaptic loss stability in the spinal cord after peripheral nerve injury [109].

As mentioned in the previous section, astrocytes expressing mSOD1 are more prone to switch to an activated inflammatory state. Díaz-Amarilla and colleagues [37] identified and isolated an astrocyte population with toxic potential towards MNs. Such population was found mainly in the spinal cord of the ALS rats during the symptomatic phase, as well as in primary astrocyte cultures. Interestingly such cells typically localized close to MNs, suggesting a link between the emergence of pathogenic cells and the rapidly progressing neurodegeneration in the rats. In another study, the astrocytic aquaporin-4 (AQP4), Cx43 and Cx30 were increased in the anterior horns of mSOD1 mice at the end stages of ALS progression [110], pointing to a deregulated cell-to-cell communication. Recently, it was demonstrated that AQP4 modulates the communication between astrocytes and microglia [111], but its effect on astrocyte inflammatory activation needs further evaluation. Actually, AQP4 has a major role in the regulation of K^+ buffering and gap junction coupling [112]. Gap junctions are necessary to form channels for direct intercellular communication, through which intracellular second messengers, such as Ca^{2+} and small molecules, are exchanged. The disruption of gap junctions between astrocytes may also account for MN degeneration and white matter pathology [113-115]. However, the role of astrocyte Cxs in brain pathologies involving neuroinflammation is rather controversial and alternates between a protective and a deleterious effect on neuronal susceptibility to damage [116].

Together, information here presented reinforces that additional knowledge on cross-talk failure between MNs and glial cells will contribute to a better understanding on how these factors act to drive/rescue ALS disease progression.

3.2. Communication Mediated by Soluble Factors

In ALS, astrocyte secretome may have a serious neurotoxic role. Several studies have reported that neurotoxins released by mSOD1 astrocytes are sufficient to decrease MN survival [74, 92, 117]. Nevertheless, the specific neurotoxic mediators that are involved remain unidentified, though different molecules have been proposed as key players in MN degeneration in ALS. Under physiological conditions, astrocytes are the glial cell type responsible for the secretion of neurotrophic factors, such as brain-derived neurotrophic factor (BDNF), glial-derived neurotrophic factor (GDNF), ciliary neurotrophic factor (CNTF) and vascular endothelial growth factor (VEGF), which are essential to the MN maintenance and function, as reviewed in [118]. However, the insufficient or deficient release of neurotrophic factors by reactive astrocytes in ALS, can promote MN damage. Moreover, astrocytes in ALS showed reduced capacity to release the metabolic substrate lactate, essential to the normal metabolism of neurons [73, 119]. Concerning excitotoxicity, astrocytes from ALS patients and mSOD1 mice were shown to release high levels of the neurotransmitter D-Serine, co-activator of NMDA receptors, favoring glutamate toxicity on MNs [120, 121]. Astrocytes isolated from the spinal cord of sALS and fALS patients also present increased expression of metabotropic glutamate receptors (mGluRs) [122]. The activation of these mGluRs promotes the secretion of Ca^{2+} and glutamate by the astrocytes, contributing to the glutamate-mediated excitotoxicity and consequent MN loss

[123]. On the other hand, the increased levels of COX-2 in mSOD1 astrocytes, as mentioned above, seem likewise to contribute to the glutamate release by the cells. COX inhibitors are however able to prevent glutamate release, reason why the use of COX-2 inhibitors was shown to improve mSOD1 mice survival [124-126].

Although little is known about the specific neuroinflammatory mediators produced by astrocytes, the analysis of gene expression in fALS and sALS astrocytes showed high expression levels of chemokines, pro-inflammatory cytokines and components of the complement pathway [74]. The ability to similarly release INFγ and to alter TGF-β signaling [117], further may induce the neuroinflammatory condition.

Regarding other neurotoxic factors, it was observed that ALS astrocytes remarkably release high levels of the pro-apoptotic mediator nerve growth factor (NGF) when stimulated by lipopolysaccharide (LPS) or peroxynitrite, thus mediating the death of vulnerable MNs [127]. Abnormal Ca^{2+} signaling in mSOD1 astrocytes, as demonstrated by Kawamata et al., [128], was shown to trigger ER Ca^{2+} overload leading to excess ER Ca^{2+} release, altered exocytosis and toxicity to MNs in culture, but also *in vivo*. This provided the first *in vivo* evidence that mSOD1 exocytosis is involved in the pathogenesis of ALS. Actually, inhibition of mSOD1 astrocytic exocytosis highly diminished toxicity toward MNs and delayed ALS progression in mSOD1 mice [128]. Therefore, taking into account the harmful effects of the astrocyte secretome in MN neurodegeneration and ALS progression, it should be considered the selective inhibition of astroglial exocytosis as a protective measure, at least at the disease onset, once the proinflammatory factors released by microglia are known to increase as the disease progresses.

3.3. Communication Mediated by Microvesicles /Exosomes

Exosome release and dissemination is another way by which astrocytes can promote neurotoxicity in MNs. Indeed, the correlation of exosomes with ALS disease was recently described [129]. Extracellular vesicles are small nanoparticles capable of transferring information between cells at long distance, being internalized by the recipient cells [130, 131]. Among them, two classes can be distinguished by their size and biogenesis mechanism. Microvesicles or ectosomes that present large diameter (100-1000 nm) and are formed by plasma membrane blebbing, and exosomes that are smaller (50-100 nm) and are released upon fusion of multivesicular bodies (MVBs) with the plasma membrane. Both extracellular vesicles can carry mRNA, miRNAs and small proteins, besides other molecules, sharing common functions in the horizontal transfer of information [36, 129]. However, miRNAs seem to be one of the the most common cargo [132]. Recent evidence indicates that both miR-155 and miR-146a, known to regulate inflammation, are released from dendritic cells within exosomes and taken up by similar cells, subsequently mediating target gene repression [133]. Moreover, this work also showed that the transfer of such miRNAs was able to modify the ability of the recipient cell to respond to an inflammatory stimulus.

Reactive astrocytes in culture were shown to release large membrane vesicles containing mitochondria, lipid droplets and ATP, under repetitive ATP stimulation [134]. Three independent works showed that mSOD1 protein may be transported by exosomes in ALS, particularly those from astrocytes [38] and from a MN-cell line (NSC-34) [31, 135, 136]. Basso et al., [38] showed that mSOD1 astrocytes from the cortical region have increased exosome release when compared with the corresponding wt cells. Moreover, the total amount of proteins associated with exosomes was significantly higher in mSOD1 astrocytes than in controls, in contrast with the lower protein levels (exosome-independent fraction) secreted in the mSOD1 astrocytes conditioned media. Authors observed that when spinal neurons were exposed to exosome-independent astrocytes conditioned media, SOD1 protein did not move into neuronal cells, while when exposed to astrocyte-derived exosomes, the spinal cord neurons indeed internalized the SOD1 protein. Importantly, the viability of MNs significantly decreased upon exposure to high concentrations of mSOD1-containing exosomes. Overall, authors showed, for the first time, that astrocyte-derived exosomes can transfer mSOD1 to neurons and induce selective MN loss, thus contributing to disease spread. Therefore, exosomes besides being a key factor in the communication between astrocytes and MNs in ALS may constitute a pathological determinant of MN degeneration. Particle size distribution by transmission electron microscopy (TEM) and by dynamic light scattering (DLS) of exosomes derived from MN-like NSC-34 cell line and astrocytes in our lab are depicted in Figure 2.

Exosomes are now recognized as an enriched source of miRNAs involved in disease dissemination, but also as having potential biomarker benefits when compared with intracellular and cell-free-blood [137, 138]. A recent study has investigated the communication between neurons and astrocytes via exosome transfer. The study evidenced that miR-124, highly expressed in the CNS, was released within neuronal exosomes and internalized by astrocytes where it upregulated, although indirectly, the expression of EAAT2 [139]. Moreover, the authors showed that SOD1G93A mice have reduced levels of miR-124 in the end-stage of the disease and that exogenous deliver of miR-124 significantly increased EAAT2 levels in such mice. It was also shown that misfolded WT SOD1 was transmitted from cell to cell through the release of protein-only aggregates or cell-derived vesicles identified as exosomes [31, 140].

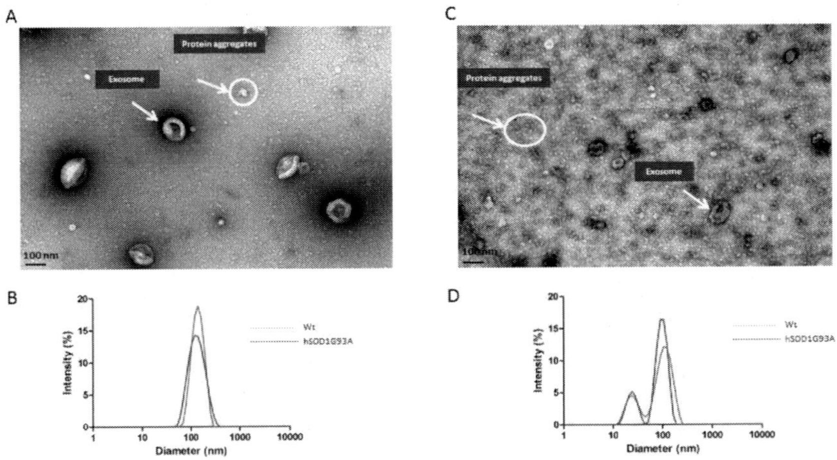

Sample	Size (diam. nm)	% Intensity
Wt	144.9	100
hSOD1G93A	141.4	100

Sample	Size (diam. nm)	% Intensity
Wt	113.3 25.5	76.1 23.9
hSOD1G93A	99.0 24.4	77.5 22.5

Figure 2. Representative images obtained by transmission electron microscopy (TEM) (A and B) and by dynamic light scattering (DLS) (C and D) of exosomes isolated from cultures of NSC-34 cells (MN-like) transfected with human SOD1G93A (mSOD1) and from cultures of astrocytes isolated from the spinal cord of mSOD1 mice, respectively. Exosomes were extracted from the extracellular media of mSOD1 MN (A, B) and astrocyte (C, D) cultures. TEM images of exosomes are depicted evidencing clusters of proteins (A, C). The evaluation of the nanoparticle (exosomes) size distribution by DLS indicated that the majority of vesicles resultant from MNs have larger diameter than those from astrocytes and located near or above 100 nm (B, D). Other small and less represented subpopulations are observed in astrocyte samples.

Consistent with previous findings relating glutamate-excitotoxicity with reduced astrocytic EAAT2, we confirmed a decreased expression of such transporter in the brain and spinal cord of early symptomatic mSOD1 mice and in isolated astrocytes from such animals, as well (unpublished data). Curiously, extracellular microvesicles from astrocytes were shown to contain functional glutamate transporters and suggested to be associated with extracellular functions [141]. Whether exosomes derived from mSOD1 astrocytes have also decreased EAAT2 content is to be evaluated.

Overall, data indicates that exosome-mediated communication exhibits a dichotomy between exosomes as key players in pathological circumstances involved in disease spreading, and as vehicles to deliver specific cargo. Whether cargo of circulating exosomes from mSOD1 astrocytes may constitute suitable biomarkers of ALS disease needs further assessment, but their presence in blood, saliva and urina [36] surely contributes to such assumption, reason why they have been proposed as targets for developing novel medicines and as upcoming therapeutic strategies.

4. Cell-Based and Exosome Delivery Therapies in ALS

After more than a century ALS has been first described, there is still no effective treatment for the disease. The only FDA approved drug Riluzole with anti-glutamatergic properties only increases patient survival by two to three months [39]. A major understanding of the upstream pathogenic mechanisms underlying ALS is still lacking which greatly inhibited the development of successful targeted therapies so far. Thus, efforts must continue to accurately define the most effective cell type, or cell types for the disease, in order to reach meaningful therapies for ALS patients. Given the evidence of ALS as a non-cell autonomous disease, therapeutics shifted its focus from neurons to non-neuronal cells. Actually, as already commented, the damage associated to astrocytes and microglia is involved in accelerated disease progression [142]. Therefore, cell-based therapy appeared as a promising strategy. Direct replacement and MN function restoration approaches using different sources of stem cells revealed to be challenging, reason why enrichment of healthy donor-derived astrocytes has been indicated to be a more realistic and beneficial approach [143].

4.1. Replacement-Based Therapies Targeting Motor Neurons

Taking into account the characteristic MN loss in ALS, the initial goal of stem cells transplantation into the spinal cord of ALS mice models was to replace impaired MNs with healthy stem-cell derived MNs, generated from neural and embryonic stem cells [144, 145]. For the generation of MNs, embryonic stem cells (ESCs) and neural stem cells (NSCs) have been used [146]. ESCs are pluripotent cells obtained from the inner cell mass of the blastocyst with risk of forming teratomas, while NSCs that are multipotent cells isolated from nervous tissue, have potential to generate neurons, astrocytes and oligodendrocytes and do not exibit such problem. Adult stem cells and iPSCs are now considered as having advantages over ESCs and NSCs, namely the immune rejection, reason why they are indicated for autologous transplantation. In such context, stem cells are a valuable way to replace the loss of spinal MNs in ALS and have been indicated to recover associated paralysis.

Transplantation of undifferentiated multipotent NSCs in a mouse model of ALS has shown to sustain vital functions such as respiration [147]. Other studies have previously revealed that transplantation of human NSCs transiently ameliorated motor performance and slightly enhanced survival in both rats and mice models of ALS [148, 149]. However, although highlighting the importance of stem cell transplantation only modest improvements were noticed in these studies [150]. Consequently, continued research in this developing area is required.

The first US FDA approved clinical trial for a stem cell-based therapy in ALS patients showed that the cell delivery using microinjections of NSCs in the spinal cord was well tolerated [151]. In a phase 1 study Feldman and coworkers [144] identified potential therapeutic windows for a better benefit of the assayed strategy and proposed the usage of several injections into both lumbar and cervical spinal cord segments. In a more recent study Mazzini and colleagues [152] used human fetal tissue specimens to generate human NSCs. A concentration of 50,000 cells/μl was tested and injected in a volume to afford a total of 750,000 cells per injection sit. The treatment was applied to 6 sALS patients (2 females and 4 males) with no adverse events. Clinical assessments, performed from 6 to 18 months after transplantation, showed no accelerated progression, but also that no significant decline was obtained. It can be assumed that the new MNs have become injured when facing the neurotoxic environment created by non-neuronal cells, since transplanted cells are expected to be as vulnerable as endogenous MNs to the toxic ALS microenvironment [146]. In this sense, new approaches towards the transplantation of cells supporting the survival of existing neurons or with the capacity to reduce the neurotoxic environment should be envisaged. In particular, it is our belief that combined therapies using transplantation of neuroprotective glial cells (engineered astrocytes), or EVs from modified donor cells, together with MN replacement will improve the therapeutic efficacy over the injection of only one type of neural cells.

iPSCs from patients with fALS are easily available to researchers at the NINDS Repository, thus allowing for the study of ALS arising from SOD1, FUS, and FIG4 mutations. However, a number of challenges regarding its usage are still remaining. Currently, the long periods of time required to generate human neural cells, namely astrocytes, results in significant expenses and often complex experimental protocols. A recent work by Meyer and collaborators [25] reported a rapid, reproducible direct conversion of adult human patient fibroblasts into tripotent iNPCs with potential to generate MNs, oligodendrocytes and, particularly, astrocytes. Because iPSCs are derived from individual patients, careful assessment of the sample sizes studied should be considered before get into broader conclusions. If used to screen for potentially relevant ALS drug targets, standardization of cell lines across laboratories will be necessary for validation of drug effects. Future aims include the development of more complex integrated structures, such as 2D and 3D models. Despite these challenges, it is likely that more efficient derivation of iPSCs will be developed, thus shortening the time and expenses currently associated with creating iPSC lines. As sources of iPSCs become more readily available to the research community, it will open the unique opportunity to explore the functional implications of a particular genotype in a cell type specific context.

4.2. Replacement-Based Therapies Targeting Glial Cells

Consistent with the important role of astrocytes in the pathophysiology of ALS, recent studies using ALS animal models proposed the targeting of astrocytic functions as a promising therapeutic avenue. Administration of the β-lactam antibiotic ceftriaxone in an ALS mouse model partially prevented the loss of EAAT2 in the spinal cord and significantly extended animal survival [153]. However, their long-term effects should be evaluated, since recent reports point out toxic effects resultant from long-term usage of ceftriaxone in rodent models [154], and loss of strength/ability to move and fatigue were reported by ALS patients [155].

Studies using cell-replacement approaches have shown that mesenchymal stem cells (MSCs) transplanted into the cisterna magna [156], or glial-restricted precursor (GPR) cells injected directly into the spinal cord [93] of mutant SOD1 rats migrated to the damaged areas, where they largely differentiated into astrocytes. Reduced microglial activation and preservation of MNs and delayed motor function decline was observed. More recently, MSCs were used as putative vehicles to deliver neuroprotective factors to CNS. Indeed, intracerebroventricular injection of encapsulated human MSCs producing glucagon-Like peptide 1 in mSOD1 mice showed neuroprotective potential by diminishing disease progression and enhancing animal survival [157]. These effects are thought to be mediated by the production and release of anti-inflammatory cytokines and chemokines [158].

Neural progenitor cells engineered to secret GDNF, promoted MN protection after transplantation into the spinal cord of mSOD1 mice, and the same cells rapidly differentiated into astrocytes replacing dying glia [159]. Another study, in which authors developed mesenchymal progenitor cells expressing a mixture of neurotrophic factors (NTFs), namely GDNF, BDNF, VEGF and Insulin-like growth factor 1 (IGF-1), a long-term delivery of NTFs was observed when cells were intramuscularly injected in the mSOD1 mice [160]. In phase 1/2 clinical trials, genetically modified MSCs to secret NTFs were transplanted by intramuscular or intrathecal injection in ALS patients and the study demonstrated that such strategy was safe and with potential clinical benefits [161].

Injection of WT astrocyte progenitors into the regions of CNS affected by ALS also constitutes a promising approach. The transplantation of rodent GRP cells into the cervical spinal cord of animal models showed their survival and differentiation into astrocytes, attenuating MN loss and extending mSOD1 rat survival [93]. Similarly, transplantation of iPSCs-derived GRP cells into the lumbar spinal cord of mSOD1 mice, besides being differentiated into astrocytes, revealed to also improve mice lifespan, mainly in male animals [94]. Transplantation of human GRP cells into mSOD1 rats and mice also showed a good survival, migration ability and astrocyte-differentiation capacity [95, 162]. However, no improvement of motor function was observed. As previously noticed for MN transplantation, the slight benefits or absent effects with the transplantation of cells differentiating into astrocytes may reside in the presence of reactive/dysfunctional astrocytes in the milieu [144]. Therefore, it is of the upmost importance to understand the mechanisms by which astrocytes acquire reactive/neurotoxic properties and to discover ways to switch the harmful phenotype into a neuroprotective cell or exosome before considering their transplantation in ALS models.

It was recently demonstrated that the fibroblast growth factor 1 (FGF1) or FGF2 promotes iPSCs-derived reactive astrocytes toward a mature quiescent phenotype [163], which may be of major relevance for modeling human pathological processes in ALS. In addition, neurons and glial cells generated from patient iPSCs, besides avoiding immune rejection, are also a very useful model to screen drugs intended to be utilized in personalized therapy, and to investigate pathological cascades [158]. Most interesting, it was demonstrated that human-derived neural progenitors into the cervical spinal cords of adult mice with severe combined immunodeficiency migrate a long distance and differentiate into astrocytes, thus providing evidence that chimeric animal models have potential to be used for investigating the role of human astrocytes in ALS pathogenesis and in repair [164].

It is expected that the use of human iPSCs and chimeric mice will increase knowledge on the ALS disease and it is anticipated that human neural progenitors from ALS patients may be engineered to deliver therapeutic compounds into the spinal cord.

4.3. Therapeutic Potential of Exosomes

To date, clinical trials with several promising compounds and using cell-transplantation techniques have not resulted in favorable outcomes in ALS patients. One of the main concerns for developing new therapies is the lack of direct translation form promising preclinical findings to successful clinical results. In spite of the deep study regarding ALS heterogeneity and complexity, early detection markers and proper biomarkers of the disease are still scarce and the current ones do not allow stratification of patients. Understanding the cellular mechanisms of protein misfolding and the way of spreading the disease are essential for a comprehensive understanding of the etiology of ALS and for a more effective therapeutic development. As previously detailed both SOD1 aggregates and SOD1-carrying exosomes are released from cells and taken up by naïve cells [31, 136]. It is certainly interesting that both vesicles and free aggregates of pathogenic proteins may be involved in the release of mSOD1 or misfolded SOD1 from affected cells and that is consistent with contiguous propagation; however, it remains to be clarified which of these mechanisms result in the effective seeding of directed misfolded content or the acquisition of toxic properties in the recipient cell.

Recently, reactive astrocytes were shown to discharge exosomes and microvesicles [134]. In any case, both mutated and misfolded SOD1 forms were shown to be efficiently transferred to MNs as protein aggregates after release from dying cells, or on exosomes secreted by living cells, causing cell death and the spread of the disease, reason why exosomes have been suggested as targets to modulate ALS disease [31, 38]. Also, it was shown that some miRNAs are specifically sorted in exosomes, whereas others are retained in cells, thus signifying the existence of mechanisms controlling the sorting into exosomes [165]. The packaging of miRNAs into vesicles is an important step to their delivery into target cells by plasma membrane fusion and endocytosis. Once they may be differently represented when released by dysfunctional cells it is conceivable that such miRNAs can be implicated as detrimental factors by significantly changing gene expression regulatory networks [166, 167]. Interestingly, exosomes may at least partially replace stem cell based-cell therapies, since it appears that a number of desired effects of MSCs are mediated via exosomes [168]. In addition, exosomes can be systemically injected to efficiently deliver miRNAs and siRNAs to target specific cells [169], including those in the brain [170]. Recently, it was evidenced that exosomes loaded with miR-155 mimic were able to significantly increase its content in cells and tissues [171]. Most promising, it was shown that NLRP-1 inflamassome proteins are present in exosomes after CNS injury and that neuronal-derived exosomes are able to cross the injured blood-spinal cord barrier, and if loaded with siRNA are able to deliver such cargo in vivo [172].

We then suggest that exosomes may be used to treat neuroinflammation, to ameliorate the dysfunctional cell homeostasis in the spinal cord and to prevent MN destruction. Strategies to promote the therapeutic application of extracellular vesicles in future clinical studies are addressed in [173].

CONCLUSION

As discussed in this book chapter, the understanding of the role of astrocytes in physiological and pathophysiological conditions, such as in neurodegenerative diseases like ALS is an expanding field of research. The contribution of astrocytes to ALS onset and progression is likely to be caused by a combination of the loss of their normal homeostasis and gain of toxic functions. Compelling evidence points to dysfunctional astrocytes in the pathogenesis of ALS. Nevertheless, the primary cause of the disease has not yet been clearly identified and other neural cell types are likely to be involved. Thus, it deserves to be investigated the effective contribution of each cell type and the molecules driving the disease, what requires the development of new technologies. For instance, the genetically modified human NSCs cultured in a 3D cell culture system were able to recapitulate key events of Alzheimer's disease pathology [174]. The recently developed 3D microtopographic scaffolds when grafted into organotypic hippocampal brain slices induced a ~3.5-fold improvement in neurite outgrowth relatively to isolated cells [175]. Therefore, these platforms that allow the delivery of multiple neuronal subtypes may represent a promising strategy for the transplantation of reprogrammed cells with neuro-regenerative capacity, such as the engineered astrocytes. The approach may be a new hope among the strategies aiming at delaying ALS progression. Indeed, there is no drug showing relevant efficacy for ALS treatment, and therapeutic benefits of spinal cord transplantation of NSCs seem to be limited to particular time frames. Neurotoxic properties of local astrocytes secreting unknown soluble neurotoxins, as well as harmful extracellular microvesicles and exosomes known to disturb normal cell-to-cell communication and disseminate the disease, are in the origin. Considering the crucial role of astrocytes in CNS homeostasis and in ALS pathology, these cells represent an interesting therapeutic target, either by re-establishing their normal signaling pathways and function, or as a vehicle to afford neuroprotection into the overall cellular environment, in particular that of MNs. Astrocyte-derived ectosomes/exosomes may also reveal to be efficient carriers of neuroprotective biocompatible cargoes to local or distant cells when delivered into the spinal cord. Based on studies revealing that miRNA content in exosomes can be manipulated, the production of exosomes from ALS patient's own cells may be an ideal complement of combined neural cell transplantation approaches.

Grant Support

Supported in part by Santa Casa da Misericórdia de Lisboa (Research Program in ALS) and Fundação para a Ciência e Tecnologia (PTDC/SAU-FAR/118787/2010, UID/DTP/04138/2013, SFRH/BPD/76590/2011, SFRH/BD/102718/2014 and SFRH/BD/91316/2012). The authors have nothing to disclose.

References

[1] Kumar DR, Aslinia F, Yale SH, Mazza JJ. Jean-Martin Charcot: the father of neurology. *Clin Med Res.* 2011;9(1):46-9.

[2] Turner MR, Brockington A, Scaber J, Hollinger H, Marsden R, Shaw PJ, et al., Pattern of spread and prognosis in lower limb-onset ALS. *Amyotroph Lateral Scler.* 2010;11(4):369-73.

[3] Kuhnlein P, Gdynia HJ, Sperfeld AD, Lindner-Pfleghar B, Ludolph AC, Prosiegel M, et al., Diagnosis and treatment of bulbar symptoms in amyotrophic lateral sclerosis. *Nat Clin Pract Neurol.* 2008;4(7):366-74.

[4] Wijesekera LC, Leigh PN. Amyotrophic lateral sclerosis. *Orphanet J Rare Dis.* 2009;4:3.

[5] Chen S, Sayana P, Zhang X, Le W. Genetics of amyotrophic lateral sclerosis: an update. *Molecular neurodegeneration.* 2013;8(1):28.

[6] Fischer LR, Culver DG, Tennant P, Davis AA, Wang M, Castellano-Sanchez A, et al., Amyotrophic lateral sclerosis is a distal axonopathy: evidence in mice and man. *Exp Neurol.* 2004;185(2):232-40.

[7] Pasinelli P, Brown RH. Molecular biology of amyotrophic lateral sclerosis: insights from genetics. *Nat Rev Neurosci.* 2006;7(9):710-23.

[8] Bendotti C, Carri MT. Lessons from models of SOD1-linked familial ALS. *Trends in molecular medicine.* 2004;10(8):393-400.

[9] Redler RL, Dokholyan NV. The complex molecular biology of amyotrophic lateral sclerosis (ALS). *Prog Mol Biol Transl Sci.* 2012;107:215-62.

[10] Proctor EA, Fee L, Tao Y, Redler RL, Fay JM, Zhang Y, et al., Nonnative SOD1 trimer is toxic to motor neurons in a model of amyotrophic lateral sclerosis. *Proc Natl Acad Sci U S A.* 2016;113(3):614-9.

[11] Shefner JM, Reaume AG, Flood DG, Scott RW, Kowall NW, Ferrante RJ, et al., Mice lacking cytosolic copper/zinc superoxide dismutase display a distinctive motor axonopathy. *Neurology.* 1999;53(6):1239-46.

[12] Gurney ME, Pu H, Chiu AY, Dal Canto MC, Polchow CY, Alexander DD, et al., Motor neuron degeneration in mice that express a human Cu,Zn superoxide dismutase mutation. *Science.* 1994;264(5166):1772-5.

[13] Gruzman A, Wood WL, Alpert E, Prasad MD, Miller RG, Rothstein JD, et al., Common molecular signature in SOD1 for both sporadic and familial amyotrophic lateral sclerosis. *Proc Natl Acad Sci U S A.* 2007;104(30):12524-9.

[14] Ezzi SA, Urushitani M, Julien JP. Wild-type superoxide dismutase acquires binding and toxic properties of ALS-linked mutant forms through oxidation. *J Neurochem.* 2007;102(1):170-8.

[15] Bosco DA, Morfini G, Karabacak NM, Song Y, Gros-Louis F, Pasinelli P, et al., Wild-type and mutant SOD1 share an aberrant conformation

and a common pathogenic pathway in ALS. *Nat Neurosci.* 2010;13(11):1396-403.

[16] Ferraiuolo L, Kirby J, Grierson AJ, Sendtner M, Shaw PJ. Molecular pathways of motor neuron injury in amyotrophic lateral sclerosis. *Nat Rev Neurol.* 2011;7(11):616-30.

[17] Baldwin KR, Godena VK, Hewitt VL, Whitworth AJ. Axonal transport defects are a common phenotype in Drosophila models of ALS. *Hum Mol Genet.* 2016.

[18] Costa J, de Carvalho M. Emerging molecular biomarker targets for amyotrophic lateral sclerosis. *Clin Chim Acta.* 2016;455:7-14.

[19] Steinacker P, Feneberg E, Weishaupt J, Brettschneider J, Tumani H, Andersen PM, et al., Neurofilaments in the diagnosis of motoneuron diseases: a prospective study on 455 patients. *J Neurol Neurosurg Psychiatry.* 2016;87(1):12-20.

[20] Menke RA, Gray E, Lu CH, Kuhle J, Talbot K, Malaspina A, et al., CSF neurofilament light chain reflects corticospinal tract degeneration in ALS. *Ann Clin Transl Neurol.* 2015;2(7):748-55.

[21] Lu CH, Macdonald-Wallis C, Gray E, Pearce N, Petzold A, Norgren N, et al., Neurofilament light chain: A prognostic biomarker in amyotrophic lateral sclerosis. *Neurology.* 2015;84(22):2247-57.

[22] Song S, Miranda CJ, Braun L, Meyer K, Frakes AE, Ferraiuolo L, et al., Major histocompatibility complex class I molecules protect motor neurons from astrocyte-induced toxicity in amyotrophic lateral sclerosis. *Nat Med.* 2016;22(4):397-403.

[23] Re DB, Le Verche V, Yu C, Amoroso MW, Politi KA, Phani S, et al., Necroptosis Drives Motor Neuron Death in Models of Both Sporadic and Familial ALS. *Neuron.* 2014;81(5):1001-8.

[24] Bilican B, Serio A, Barmada SJ, Nishimura AL, Sullivan GJ, Carrasco M, et al., Mutant induced pluripotent stem cell lines recapitulate aspects of TDP-43 proteinopathies and reveal cell-specific vulnerability. *Proc Natl Acad Sci U S A.* 2012;109(15):5803-8.

[25] Meyer K, Ferraiuolo L, Miranda CJ, Likhite S, McElroy S, Renusch S, et al., Direct conversion of patient fibroblasts demonstrates non-cell autonomous toxicity of astrocytes to motor neurons in familial and sporadic ALS. *Proc Natl Acad Sci U S A.* 2014;111(2):829-32.

[26] Boillée S, Vande Velde C, Cleveland DW. ALS: a disease of motor neurons and their nonneuronal neighbors. *Neuron.* 2006;52(1):39-59.

[27] Pramatarova A, Laganiere J, Roussel J, Brisebois K, Rouleau GA. Neuron-specific expression of mutant superoxide dismutase 1 in

transgenic mice does not lead to motor impairment. *J Neurosci.* 2001;21(10):3369-74.
[28] Gong YH, Parsadanian AS, Andreeva A, Snider WD, Elliott JL. Restricted expression of G86R Cu/Zn superoxide dismutase in astrocytes results in astrocytosis but does not cause motoneuron degeneration. *J Neurosci.* 2000;20(2):660-5.
[29] Boillée S, Yamanaka K, Lobsiger CS, Copeland NG, Jenkins NA, Kassiotis G, et al., Onset and progression in inherited ALS determined by motor neurons and microglia. *Science.* 2006;312(5778):1389-92.
[30] Yamanaka K, Chun SJ, Boillee S, Fujimori-Tonou N, Yamashita H, Gutmann DH, et al., Astrocytes as determinants of disease progression in inherited amyotrophic lateral sclerosis. *Nat Neurosci.* 2008;11(3):251-3.
[31] Grad LI, Yerbury JJ, Turner BJ, Guest WC, Pokrishevsky E, O'Neill MA, et al., Intercellular propagated misfolding of wild-type Cu/Zn superoxide dismutase occurs via exosome-dependent and -independent mechanisms. *Proc Natl Acad Sci U S A.* 2014;111(9):3620-5.
[32] Vardjan N, Zorec R. Excitable Astrocytes: Ca(2+)- and cAMP-Regulated Exocytosis. *Neurochem Res.* 2015;40(12):2414-24.
[33] Verkhratsky A, Matteoli M, Parpura V, Mothet JP, Zorec R. Astrocytes as secretory cells of the central nervous system: idiosyncrasies of vesicular secretion. *EMBO J.* 2016;35(3):239-57.
[34] Pekny M, Pekna M, Messing A, Steinhauser C, Lee JM, Parpura V, et al., Astrocytes: a central element in neurological diseases. *Acta Neuropathol.* 2016;131(3):323-45.
[35] Choi DS, Kim DK, Kim YK, Gho YS. Proteomics, transcriptomics and lipidomics of exosomes and ectosomes. *Proteomics.* 2013;13(10-11):1554-71.
[36] Brites D, Fernandes A. Neuroinflammation and Depression: Microglia Activation, Extracellular Microvesicles and microRNA Dysregulation. *Front Cell Neurosci.* 2015;9:476.
[37] Díaz-Amarilla P, Olivera-Bravo S, Trias E, Cragnolini A, Martinez-Palma L, Cassina P, et al., Phenotypically aberrant astrocytes that promote motoneuron damage in a model of inherited amyotrophic lateral sclerosis. *Proc Natl Acad Sci U S A.* 2011;108(44):18126-31.
[38] Basso M, Pozzi S, Tortarolo M, Fiordaliso F, Bisighini C, Pasetto L, et al., Mutant copper-zinc superoxide dismutase (SOD1) induces protein secretion pathway alterations and exosome release in astrocytes:

implications for disease spreading and motor neuron pathology in amyotrophic lateral sclerosis. *J Biol Chem.* 2013;288(22):15699-711.

[39] Ludolph AC, Jesse S. Evidence-based drug treatment in amyotrophic lateral sclerosis and upcoming clinical trials. *Ther Adv Neurol Disord.* 2009;2(5):319-26.

[40] Chung WS, Barres BA. The role of glial cells in synapse elimination. *Curr Opin Neurobiol.* 2012;22(3):438-45.

[41] Allaman I, Belanger M, Magistretti PJ. Astrocyte-neuron metabolic relationships: for better and for worse. *Trends Neurosci.* 2011;34(2):76-87.

[42] Sofroniew MV, Vinters HV. Astrocytes: biology and pathology. *Acta Neuropathol.* 2010;119(1):7-35.

[43] Simon NG, Turner MR, Vucic S, Al-Chalabi A, Shefner J, Lomen-Hoerth C, et al., Quantifying disease progression in amyotrophic lateral sclerosis. *Ann Neurol.* 2014;76(5):643-57.

[44] Andersen PM, Al-Chalabi A. Clinical genetics of amyotrophic lateral sclerosis: what do we really know? *Nat Rev Neurol.* 2011;7(11):603-15.

[45] Rakhit R, Crow JP, Lepock JR, Kondejewski LH, Cashman NR, Chakrabartty A. Monomeric Cu,Zn-superoxide dismutase is a common misfolding intermediate in the oxidation models of sporadic and familial amyotrophic lateral sclerosis. *J Biol Chem.* 2004;279(15):15499-504.

[46] McCombe PA, Henderson RD. The Role of immune and inflammatory mechanisms in ALS. *Curr Mol Med.* 2011;11(3):246-54.

[47] Philips T, Robberecht W. Neuroinflammation in amyotrophic lateral sclerosis: role of glial activation in motor neuron disease. *Lancet Neurol.* 2011;10(3):253-63.

[48] Saleh IA, Zesiewicz T, Xie Y, Sullivan KL, Miller AM, Kuzmin-Nichols N, et al., Evaluation of humoral immune response in adaptive immunity in ALS patients during disease progression. *J Neuroimmunol.* 2009;215(1-2):96-101.

[49] Migheli A, Cordera S, Bendotti C, Atzori C, Piva R, Schiffer D. S-100beta protein is upregulated in astrocytes and motor neurons in the spinal cord of patients with amyotrophic lateral sclerosis. *Neurosci Lett.* 1999;261(1-2):25-8.

[50] Süssmuth SD, Sperfeld AD, Hinz A, Brettschneider J, Endruhn S, Ludolph AC, et al., CSF glial markers correlate with survival in amyotrophic lateral sclerosis. *Neurology.* 2010;74(12):982-7.

[51] Süssmuth SD, Tumani H, Ecker D, Ludolph AC. Amyotrophic lateral sclerosis: disease stage related changes of tau protein and S100 beta in

cerebrospinal fluid and creatine kinase in serum. *Neurosci Lett.* 2003;353(1):57-60.

[52] Hoyaux D, Boom A, Van den Bosch L, Belot N, Martin JJ, Heizmann CW, et al., S100A6 overexpression within astrocytes associated with impaired axons from both ALS mouse model and human patients. *J Neuropathol Exp Neurol.* 2002;61(8):736-44.

[53] Kaplan A, Spiller KJ, Towne C, Kanning KC, Choe GT, Geber A, et al., Neuronal matrix metalloproteinase-9 is a determinant of selective neurodegeneration. *Neuron.* 2014;81(2):333-48.

[54] Vaz AR, Cunha C, Gomes C, Schmucki N, Barbosa M, Brites D. Glycoursodeoxycholic acid reduces matrix metalloproteinase-9 and caspase-9 activation in a cellular model of superoxide dismutase-1 neurodegeneration. *Mol Neurobiol.* 2015;51(3):864-77.

[55] Demestre M, Parkin-Smith G, Petzold A, Pullen AH. The pro and the active form of matrix metalloproteinase-9 is increased in serum of patients with amyotrophic lateral sclerosis. *J Neuroimmunol.* 2005;159(1-2):146-54.

[56] Patin F, Corcia P, Madji Hounoum B, Veyrat-Durebex C, Respaud E, Piver E, et al., Biological follow-up in amyotrophic lateral sclerosis: decrease in creatinine levels and increase in ferritin levels predict poor prognosis. *Eur J Neurol.* 2015; 22(10):1385-90.

[57] Shibata N, Hirano A, Hedley-Whyte ET, Dal Canto MC, Nagai R, Uchida K, et al., Selective formation of certain advanced glycation end products in spinal cord astrocytes of humans and mice with superoxide dismutase-1 mutation. *Acta Neuropathol.* 2002;104(2):171-8.

[58] Ilzecka J. Serum-soluble receptor for advanced glycation end product levels in patients with amyotrophic lateral sclerosis. *Acta Neurol Scand.* 2009;120(2):119-22.

[59] Boison D, Aronica E. Comorbidities in Neurology: Is adenosine the common link? *Neuropharmacology.* 2015;97:18-34.

[60] Almad AA, Doreswamy A, Gross SK, Richard JP, Huo Y, Haughey N, et al., Connexin 43 in astrocytes contributes to motor neuron toxicity in amyotrophic lateral sclerosis. *Glia.* 2016.

[61] Rothstein JD, Tsai G, Kuncl RW, Clawson L, Cornblath DR, Drachman DB, et al., Abnormal excitatory amino acid metabolism in amyotrophic lateral sclerosis. *Ann Neurol.* 1990;28(1):18-25.

[62] Texido L, Hernandez S, Martin-Satue M, Povedano M, Casanovas A, Esquerda J, et al., Sera from amyotrophic lateral sclerosis patients

induce the non-canonical activation of NMDA receptors "in vitro." *Neurochem Int.* 2011;59(6):954-64.

[63] Amara SG, Fontana AC. Excitatory amino acid transporters: keeping up with glutamate. *Neurochem Int.* 2002;41(5):313-8.

[64] Bristol LA, Rothstein JD. Glutamate transporter gene expression in amyotrophic lateral sclerosis motor cortex. *Ann Neurol.* 1996;39(5):676-9.

[65] Howland DS, Liu J, She Y, Goad B, Maragakis NJ, Kim B, et al., Focal loss of the glutamate transporter EAAT2 in a transgenic rat model of SOD1 mutant-mediated amyotrophic lateral sclerosis (ALS). *Proc Natl Acad Sci U S A.* 2002;99(3):1604-9.

[66] Barbeito LH, Pehar M, Cassina P, Vargas MR, Peluffo H, Viera L, et al., A role for astrocytes in motor neuron loss in amyotrophic lateral sclerosis. *Brain Res Brain Res Rev.* 2004;47(1-3):263-74.

[67] Guo Y, Duan W, Li Z, Huang J, Yin Y, Zhang K, et al., Decreased GLT-1 and increased SOD1 and HO-1 expression in astrocytes contribute to lumbar spinal cord vulnerability of SOD1-G93A transgenic mice. *FEBS Lett.* 2010;584(8):1615-22.

[68] Rao SD, Yin HZ, Weiss JH. Disruption of glial glutamate transport by reactive oxygen species produced in motor neurons. *J Neurosci.* 2003;23(7):2627-33.

[69] Li K, Hala TJ, Seetharam S, Poulsen DJ, Wright MC, Lepore AC. GLT1 overexpression in SOD1(G93A) mouse cervical spinal cord does not preserve diaphragm function or extend disease. *Neurobiol Dis.* 2015;78:12-23.

[70] Van Damme P, Bogaert E, Dewil M, Hersmus N, Kiraly D, Scheveneels W, et al., Astrocytes regulate GluR2 expression in motor neurons and their vulnerability to excitotoxicity. *Proc Natl Acad Sci U S A.* 2007;104(37):14825-30.

[71] Almer G, Vukosavic S, Romero N, Przedborski S. Inducible nitric oxide synthase up-regulation in a transgenic mouse model of familial amyotrophic lateral sclerosis. *J Neurochem.* 1999;72(6):2415-25.

[72] Sasaki S, Warita H, Abe K, Iwata M. Inducible nitric oxide synthase (iNOS) and nitrotyrosine immunoreactivity in the spinal cords of transgenic mice with a G93A mutant SOD1 gene. *J Neuropathol Exp Neurol.* 2001;60(9):839-46.

[73] Cassina P, Cassina A, Pehar M, Castellanos R, Gandelman M, de Leon A, et al., Mitochondrial dysfunction in SOD1G93A-bearing astrocytes

promotes motor neuron degeneration: prevention by mitochondrial-targeted antioxidants. *J Neurosci.* 2008;28(16):4115-22.

[74] Haidet-Phillips AM, Hester ME, Miranda CJ, Meyer K, Braun L, Frakes A, et al., Astrocytes from familial and sporadic ALS patients are toxic to motor neurons. *Nat Biotechnol.* 2011;29(9):824-8.

[75] Endo F, Komine O, Fujimori-Tonou N, Katsuno M, Jin S, Watanabe S, et al., Astrocyte-derived TGF-beta1 accelerates disease progression in ALS mice by interfering with the neuroprotective functions of microglia and T cells. *Cell Rep.* 2015;11(4):592-604.

[76] Hensley K, Abdel-Moaty H, Hunter J, Mhatre M, Mou S, Nguyen K, et al., Primary glia expressing the G93A-SOD1 mutation present a neuroinflammatory phenotype and provide a cellular system for studies of glial inflammation. *J Neuroinflammation.* 2006;3:2.

[77] Hensley K, Mhatre M, Mou S, Pye QN, Stewart C, West M, et al., On the relation of oxidative stress to neuroinflammation: lessons learned from the G93A-SOD1 mouse model of amyotrophic lateral sclerosis. *Antioxid Redox Signal.* 2006;8(11-12):2075-87.

[78] Johann S, Heitzer M, Kanagaratnam M, Goswami A, Rizo T, Weis J, et al., NLRP3 inflammasome is expressed by astrocytes in the SOD1 mouse model of ALS and in human sporadic ALS patients. *Glia.* 2015;63(12):2260-73.

[79] Guo H, Callaway JB, Ting JP. Inflammasomes: mechanism of action, role in disease, and therapeutics. *Nat Med.* 2015;21(7):677-87.

[80] Tomankova T, Petrek M, Gallo J, Kriegova E. MicroRNAs: Emerging Regulators of Immune-Mediated Diseases. *Scand J Immunol.* 2011;75(2):129-41.

[81] 8Kye MJ, Goncalves Ido C. The role of miRNA in motor neuron disease. *Front Cell Neurosci.* 2014;8:15.

[82] Krol J, Loedige I, Filipowicz W. The widespread regulation of microRNA biogenesis, function and decay. *Nat Rev Genet.* 2010;11(9):597-610.

[83] Parisi C, Arisi I, D'Ambrosi N, Storti AE, Brandi R, D'Onofrio M, et al., Dysregulated microRNAs in amyotrophic lateral sclerosis microglia modulate genes linked to neuroinflammation. *Cell Death Dis.* 2013;4:e959.

[84] Volonté C, Apolloni S, Parisi C. MicroRNAs: newcomers into the ALS picture. *CNS Neurol Disord Drug Targets.* 2015;14(2):194-207.

[85] Figueroa-Romero C, Hur J, Lunn JS, Paez-Colasante X, Bender DE, Yung R, et al., Expression of microRNAs in human post-mortem

amyotrophic lateral sclerosis spinal cords provides insight into disease mechanisms. *Mol Cell Neurosci.* 2015;71:34-45.

[86] Koval ED, Shaner C, Zhang P, du Maine X, Fischer K, Tay J, et al., Method for widespread microRNA-155 inhibition prolongs survival in ALS-model mice. *Hum Mol Genet.* 2013;22(20):4127-35.

[87] Butovsky O, Jedrychowski MP, Cialic R, Krasemann S, Murugaiyan G, Fanek Z, et al., Targeting miR-155 restores abnormal microglia and attenuates disease in SOD1 mice. *Ann Neurol.* 2014;77(1):75-99.

[88] Magna M, Pisetsky DS. The role of HMGB1 in the pathogenesis of inflammatory and autoimmune diseases. *Mol Med.* 2014;20:138-46.

[89] Smith TW, Tyler HR, Schoene WC. Atypical astrocytes and Rosenthal fibers in a case of amyotrophic lateral sclerosis associated with a cerebral glioblastoma multiforme. *Acta Neuropathol.* 1975;31(1):29-34.

[90] Kamo H, Haebara H, Akiguchi I, Kameyama M, Kimura H, McGeer PL. A distinctive distribution of reactive astroglia in the precentral cortex in amyotrophic lateral sclerosis. *Acta Neuropathol.* 1987;74(1):33-8.

[91] Nagy D, Kato T, Kushner PD. Reactive astrocytes are widespread in the cortical gray matter of amyotrophic lateral sclerosis. *J Neurosci Res.* 1994;38(3):336-47.

[92] Nagai M, Re DB, Nagata T, Chalazonitis A, Jessell TM, Wichterle H, et al., Astrocytes expressing ALS-linked mutated SOD1 release factors selectively toxic to motor neurons. *Nat Neurosci.* 2007;10(5):615-22.

[93] Lepore AC, Rauck B, Dejea C, Pardo AC, Rao MS, Rothstein JD, et al., Focal transplantation-based astrocyte replacement is neuroprotective in a model of motor neuron disease. *Nat Neurosci.* 2008;11(11):1294-301.

[94] Kondo T, Funayama M, Tsukita K, Hotta A, Yasuda A, Nori S, et al., Focal transplantation of human iPSC-derived glial-rich neural progenitors improves lifespan of ALS mice. *Stem Cell Reports.* 2014;3(2):242-9.

[95] Lepore AC, O'Donnell J, Kim AS, Williams T, Tuteja A, Rao MS, et al., Human glial-restricted progenitor transplantation into cervical spinal cord of the SOD1 mouse model of ALS. *PLoS One.* 2011;6(10):e25968.

[96] Brites D, Vaz AR. Microglia centered pathogenesis in ALS: insights in cell interconnectivity. *Front Cell Neurosci.* 2014;8:117.

[97] Khakh BS, Sofroniew MV. Diversity of astrocyte functions and phenotypes in neural circuits. *Nat Neurosci.* 2015;18(7):942-52.

[98] Tarassishin L, Suh HS, Lee SC. LPS and IL-1 differentially activate mouse and human astrocytes: role of CD14. *Glia.* 2014;62(6):999-1013.

[99] Nicaise C, Mitrecic D, Falnikar A, Lepore AC. Transplantation of stem cell-derived astrocytes for the treatment of amyotrophic lateral sclerosis and spinal cord injury. *World J Stem Cells*. 2015;7(2):380-98.

[100] Di Giorgio FP, Carrasco MA, Siao MC, Maniatis T, Eggan K. Non-cell autonomous effect of glia on motor neurons in an embryonic stem cell-based ALS model. *Nat Neurosci.* 2007;10(5):608-14.

[101] Di Giorgio FP, Boulting GL, Bobrowicz S, Eggan KC. Human embryonic stem cell-derived motor neurons are sensitive to the toxic effect of glial cells carrying an ALS-causing mutation. *Cell stem cell.* 2008;3(6):637-48.

[102] Marchetto MC, Muotri AR, Mu Y, Smith AM, Cezar GG, Gage FH. Non-cell-autonomous effect of human SOD1 G37R astrocytes on motor neurons derived from human embryonic stem cells. *Cell Stem Cell.* 2008;3(6):649-57.

[103] Koizumi S, Fujishita K, Inoue K. Regulation of cell-to-cell communication mediated by astrocytic ATP in the CNS. *Purinergic Signal.* 2005;1(3):211-7.

[104] Fellin T, Sul JY, D'Ascenzo M, Takano H, Pascual O, Haydon PG. Bidirectional astrocyte-neuron communication: the many roles of glutamate and ATP. Novartis Found Symp. 2006;276:208-17; *discussion* 17-21, 33-7, 75-81.

[105] Yang C, Wang H, Qiao T, Yang B, Aliaga L, Qiu L, et al., Partial loss of TDP-43 function causes phenotypes of amyotrophic lateral sclerosis. *Proc Natl Acad Sci U S A.* 2014;111(12):E1121-9.

[106] Rojas F, Cortes N, Abarzua S, Dyrda A, van Zundert B. Astrocytes expressing mutant SOD1 and TDP43 trigger motoneuron death that is mediated via sodium channels and nitroxidative stress. *Front Cell Neurosci.* 2014;8:24.

[107] Casula M, Iyer AM, Spliet WG, Anink JJ, Steentjes K, Sta M, et al., Toll-like receptor signaling in amyotrophic lateral sclerosis spinal cord tissue. *Neuroscience.* 2011;179:233-43.

[108] Okun E, Griffioen KJ, Son TG, Lee JH, Roberts NJ, Mughal MR, et al., TLR2 activation inhibits embryonic neural progenitor cell proliferation. *J Neurochem*. 2010;114(2):462-74.

[109] Freria CM, Velloso LA, Oliveira AL. Opposing effects of Toll-like receptors 2 and 4 on synaptic stability in the spinal cord after peripheral nerve injury. *J Neuroinflammation*. 2012;9:240.

[110] Cui Y, Masaki K, Yamasaki R, Imamura S, Suzuki SO, Hayashi S, et al., Extensive dysregulations of oligodendrocytic and astrocytic connexins

are associated with disease progression in an amyotrophic lateral sclerosis mouse model. *J Neuroinflammation*. 2014;11:42.

[111] Sun H, Liang R, Yang B, Zhou Y, Liu M, Fang F, et al., Aquaporin-4 mediates communication between astrocyte and microglia: Implications of neuroinflammation in experimental Parkinson's disease. *Neuroscience*. 2016;317:65-75.

[112] Strohschein S, Huttmann K, Gabriel S, Binder DK, Heinemann U, Steinhauser C. Impact of aquaporin-4 channels on K+ buffering and gap junction coupling in the hippocampus. *Glia*. 2011;59(6):973-80.

[113] Takeuchi H, Suzumura A. Gap junctions and hemichannels composed of connexins: potential therapeutic targets for neurodegenerative diseases. *Front Cell Neurosci*. 2014;8:189.

[114] Nakase T, Naus CC. Gap junctions and neurological disorders of the central nervous system. *Biochim Biophys Acta*. 2004;1662(1-2):149-58.

[115] Lutz SE, Zhao Y, Gulinello M, Lee SC, Raine CS, Brosnan CF. Deletion of astrocyte connexins 43 and 30 leads to a dysmyelinating phenotype and hippocampal CA1 vacuolation. *J Neurosci*. 2009;29(24):7743-52.

[116] Retamal MA, Froger N, Palacios-Prado N, Ezan P, Saez PJ, Saez JC, et al., Cx43 hemichannels and gap junction channels in astrocytes are regulated oppositely by proinflammatory cytokines released from activated microglia. *J Neurosci*. 2007;27(50):13781-92.

[117] Valori CF, Brambilla L, Martorana F, Rossi D. The multifaceted role of glial cells in amyotrophic lateral sclerosis. *Cell Mol Life Sci*. 2014;71(2):287-97.

[118] Staats KA, Van Den Bosch L. Astrocytes in amyotrophic lateral sclerosis: direct effects on motor neuron survival. *J Biol Phys*. 2009;35(4):337-46.

[119] Ferraiuolo L, Higginbottom A, Heath PR, Barber S, Greenald D, Kirby J, et al., Dysregulation of astrocyte-motoneuron cross-talk in mutant superoxide dismutase 1-related amyotrophic lateral sclerosis. *Brain*. 2011;134(Pt 9):2627-41.

[120] Sasabe J, Chiba T, Yamada M, Okamoto K, Nishimoto I, Matsuoka M, et al., D-serine is a key determinant of glutamate toxicity in amyotrophic lateral sclerosis. *EMBO J*. 2007;26(18):4149-59.

[121] Sasabe J, Miyoshi Y, Suzuki M, Mita M, Konno R, Matsuoka M, et al., D-amino acid oxidase controls motoneuron degeneration through D-serine. *Proc Natl Acad Sci U S A*. 2012;109(2):627-32.

[122] Aronica E, Catania MV, Geurts J, Yankaya B, Troost D. Immunohistochemical localization of group I and II metabotropic

glutamate receptors in control and amyotrophic lateral sclerosis human spinal cord: upregulation in reactive astrocytes. *Neuroscience.* 2001;105(2):509-20.

[123] Seifert G, Schilling K, Steinhauser C. Astrocyte dysfunction in neurological disorders: a molecular perspective. *Nat Rev Neurosci.* 2006;7(3):194-206.

[124] Klivenyi P, Kiaei M, Gardian G, Calingasan NY, Beal MF. Additive neuroprotective effects of creatine and cyclooxygenase 2 inhibitors in a transgenic mouse model of amyotrophic lateral sclerosis. *J Neurochem.* 2004;88(3):576-82.

[125] Drachman DB, Frank K, Dykes-Hoberg M, Teismann P, Almer G, Przedborski S, et al., Cyclooxygenase 2 inhibition protects motor neurons and prolongs survival in a transgenic mouse model of ALS. *Ann Neurol.* 2002;52(6):771-8.

[126] Pompl PN, Ho L, Bianchi M, McManus T, Qin W, Pasinetti GM. A therapeutic role for cyclooxygenase-2 inhibitors in a transgenic mouse model of amyotrophic lateral sclerosis. *FASEB J.* 2003;17(6):725-7.

[127] Pehar M, Cassina P, Vargas MR, Castellanos R, Viera L, Beckman JS, et al., Astrocytic production of nerve growth factor in motor neuron apoptosis: implications for amyotrophic lateral sclerosis. *J Neurochem.* 2004;89(2):464-73.

[128] Kawamata H, Ng SK, Diaz N, Burstein S, Morel L, Osgood A, et al., Abnormal intracellular calcium signaling and SNARE-dependent exocytosis contributes to SOD1G93A astrocyte-mediated toxicity in amyotrophic lateral sclerosis. *J Neurosci.* 2014;34(6):2331-48.

[129] Aryani A, Denecke B. Exosomes as a Nanodelivery System: a Key to the Future of Neuromedicine? *Mol Neurobiol.* 2016;53(2):818-34.

[130] Lee Y, El Andaloussi S, Wood MJ. Exosomes and microvesicles: extracellular vesicles for genetic information transfer and gene therapy. *Hum Mol Genet.* 2012;21(R1):R125-34.

[131] Feng D, Zhao WL, Ye YY, Bai XC, Liu RQ, Chang LF, et al., Cellular internalization of exosomes occurs through phagocytosis. *Traffic.* 2010;11(5):675-87.

[132] Chen TS, Lai RC, Lee MM, Choo AB, Lee CN, Lim SK. Mesenchymal stem cell secretes microparticles enriched in pre-microRNAs. *Nucleic Acids Res.* 2010;38(1):215-24.

[133] Alexander M, Hu R, Runtsch MC, Kagele DA, Mosbruger TL, Tolmachova T, et al., Exosome-delivered microRNAs modulate the inflammatory response to endotoxin. *Nat Commun.* 2015;6:7321.

[134] Falchi AM, Sogos V, Saba F, Piras M, Congiu T, Piludu M. Astrocytes shed large membrane vesicles that contain mitochondria, lipid droplets and ATP. *Histochem Cell Biol.* 2013;139(2):221-31.

[135] Gomes C, Keller S, Altevogt P, Costa J. Evidence for secretion of Cu,Zn superoxide dismutase via exosomes from a cell model of amyotrophic lateral sclerosis. *Neurosci Lett.* 2007;428(1):43-6.

[136] Silverman JM, Fernando SM, Grad LI, Hill AF, Turner BJ, Yerbury JJ, et al., Disease Mechanisms in ALS: Misfolded SOD1 Transferred Through Exosome-Dependent and Exosome-Independent Pathways. *Cell Mol Neurobio*l. 2016;36(3):377-81.

[137] Cheng L, Sharples RA, Scicluna BJ, Hill AF. Exosomes provide a protective and enriched source of miRNA for biomarker profiling compared to intracellular and cell-free blood. *J Extracell Vesicles.* 2014;3.

[138] Schneider A, Simons M. Exosomes: vesicular carriers for intercellular communication in neurodegenerative disorders. *Cell Tissue Res.* 2013;352(1):33-47.

[139] Morel L, Regan M, Higashimori H, Ng SK, Esau C, Vidensky S, et al., Neuronal exosomal miRNA-dependent translational regulation of astroglial glutamate transporter GLT1. *J Biol Chem.* 2013;288(10):7105-16.

[140] Grad LI, Pokrishevsky E, Silverman JM, Cashman NR. Exosome-dependent and independent mechanisms are involved in prion-like transmission of propagated Cu/Zn superoxide dismutase misfolding. *Prion.* 2014;8(5):331-5.

[141] Gosselin RD, Meylan P, Decosterd I. Extracellular microvesicles from astrocytes contain functional glutamate transporters: regulation by protein kinase C and cell activation. *Front Cell Neurosci.* 2013;7:251.

[142] Lasiene J, Yamanaka K. Glial cells in amyotrophic lateral sclerosis. *Neurol Res Int.* 2011;2011:718987.

[143] Kruminis-Kaszkiel E, Wojtkiewicz J, Maksymowicz W. Glial-restricted precursors as potential candidates for ALS cell-replacement therapy. *Acta Neurobiol Exp (Wars).* 2014;74(3):233-41.

[144] Feldman EL, Boulis NM, Hur J, Johe K, Rutkove SB, Federici T, et al., Intraspinal neural stem cell transplantation in amyotrophic lateral sclerosis: phase 1 trial outcomes. *Ann Neurol.* 2014;75(3):363-73.

[145] Lopez-Gonzalez R, Kunckles P, Velasco I. Transient recovery in a rat model of familial amyotrophic lateral sclerosis after transplantation of

motor neurons derived from mouse embryonic stem cells. *Cell Transplant.* 2009;18(10):1171-81.

[146] Thonhoff JR, Ojeda L, Wu P. Stem cell-derived motor neurons: applications and challenges in amyotrophic lateral sclerosis. *Curr Stem Cell Res Ther.* 2009;4(3):178-99.

[147] Teng YD, Benn SC, Kalkanis SN, Shefner JM, Onario RC, Cheng B, et al., Multimodal actions of neural stem cells in a mouse model of ALS: a meta-analysis. *Sci Transl Med.* 2012;4(165):165ra4.

[148] Xu L, Yan J, Chen D, Welsh AM, Hazel T, Johe K, et al., Human neural stem cell grafts ameliorate motor neuron disease in SOD-1 transgenic rats. *Transplantation.* 2006;82(7):865-75.

[149] Yan J, Xu L, Welsh AM, Chen D, Hazel T, Johe K, et al., Combined immunosuppressive agents or CD4 antibodies prolong survival of human neural stem cell grafts and improve disease outcomes in amyotrophic lateral sclerosis transgenic mice. *Stem Cells.* 2006;24(8):1976-85.

[150] Ferraiuolo L, Frakes A, Kaspar BK. Neural stem cells as a therapeutic approach for amyotrophic lateral sclerosis. *Mol Ther.* 2013;21(3):503-5.

[151] Riley J, Glass J, Feldman EL, Polak M, Bordeau J, Federici T, et al., Intraspinal stem cell transplantation in amyotrophic lateral sclerosis: a phase I trial, cervical microinjection, and final surgical safety outcomes. *Neurosurgery.* 2014;74(1):77-87.

[152] Mazzini L, Gelati M, Profico DC, Sgaravizzi G, Projetti Pensi M, Muzi G, et al., Human neural stem cell transplantation in ALS: initial results from a phase I trial. *J Transl Med.* 2015;13:17.

[153] Rothstein JD, Patel S, Regan MR, Haenggeli C, Huang YH, Bergles DE, et al., Beta-lactam antibiotics offer neuroprotection by increasing glutamate transporter expression. *Nature.* 2005;433(7021):73-7.

[154] Ratti E, Berry JD, Greenblatt DJ, Loci L, Ellrodt AS, Shefner JM, et al., Preclinical Rodent Toxicity Studies for Long Term Use of Ceftriaxone. *Toxicology reports.* 2015;2:1396-403.

[155] Raheja D, Stephens HE, Lehman E, Walsh S, Yang C, Simmons Z. Patient-reported problematic symptoms in an ALS treatment trial. *Amyotrophic lateral sclerosis & frontotemporal degeneration.* 2016:1-8.

[156] Boucherie C, Schafer S, Lavand'homme P, Maloteaux JM, Hermans E. Chimerization of astroglial population in the lumbar spinal cord after mesenchymal stem cell transplantation prolongs survival in a rat model of amyotrophic lateral sclerosis. *J Neurosci Res.* 2009;87(9):2034-46.

[157] Knippenberg S, Thau N, Dengler R, Brinker T, Petri S. Intracerebroventricular injection of encapsulated human mesenchymal

cells producing glucagon-like peptide 1 prolongs survival in a mouse model of ALS. *PLoS One.* 2012;7(6):e36857.

[158] Lunn JS, Sakowski SA, Feldman EL. Concise review: Stem cell therapies for amyotrophic lateral sclerosis: recent advances and prospects for the future. *Stem cells.* 2014;32(5):1099-109.

[159] Suzuki M, McHugh J, Tork C, Shelley B, Klein SM, Aebischer P, et al., GDNF secreting human neural progenitor cells protect dying motor neurons, but not their projection to muscle, in a rat model of familial ALS. *PLoS One.* 2007;2(8):e689.

[160] Dadon-Nachum M, Ben-Yaacov K, Ben-Zur T, Barhum Y, Yaffe D, Perlson E, et al., Transplanted modified muscle progenitor cells expressing a mixture of neurotrophic factors delay disease onset and enhance survival in the SOD1 mouse model of ALS. *J Mol Neurosci.* 2015;55(3):788-97.

[161] Petrou P, Gothelf Y, Argov Z, Gotkine M, Levy YS, Kassis I, et al., Safety and Clinical Effects of Mesenchymal Stem Cells Secreting Neurotrophic Factor Transplantation in Patients With Amyotrophic Lateral Sclerosis: Results of Phase 1/2 and 2a Clinical Trials. *JAMA neurology.* 2016;73(3):337-44.

[162] Haidet-Phillips AM, Doreswamy A, Gross SK, Tang X, Campanelli JT, Maragakis NJ. Human glial progenitor engraftment and gene expression is independent of the ALS environment. *Exp Neurol.* 2015;264:188-99.

[163] Roybon L, Lamas NJ, Garcia-Diaz A, Yang EJ, Sattler R, Jackson-Lewis V, et al., Human stem cell-derived spinal cord astrocytes with defined mature or reactive phenotypes. *Cell Rep.* 2013;4(5):1035-48.

[164] Chen H, Qian K, Chen W, Hu B, Blackbourn LWt, Du Z, et al., Human-derived neural progenitors functionally replace astrocytes in adult mice. *J Clin Invest.* 2015;125(3):1033-42.

[165] Villarroya-Beltri C, Gutierrez-Vazquez C, Sanchez-Cabo F, Perez-Hernandez D, Vazquez J, Martin-Cofreces N, et al., Sumoylated hnRNPA2B1 controls the sorting of miRNAs into exosomes through binding to specific motifs. *Nat Commun.* 2013;4:2980.

[166] Jung HJ, Suh Y. MicroRNA in Aging: From Discovery to Biology. *Curr Genomics.* 2012;13(7):548-57.

[167] Weilner S, Schraml E, Redl H, Grillari-Voglauer R, Grillari J. Secretion of microvesicular miRNAs in cellular and organismal aging. *Exp Gerontol.* 2013;48(7):626-33.

[168] Braccioli L, van Velthoven C, Heijnen CJ. Exosomes: a new weapon to treat the central nervous system. *Mol Neurobiol.* 2014;49(1):113-9.

[169] Ohno S, Takanashi M, Sudo K, Ueda S, Ishikawa A, Matsuyama N, et al., Systemically injected exosomes targeted to EGFR deliver antitumor microRNA to breast cancer cells. *Mol Ther.* 2013;21(1):185-91.

[170] Alvarez-Erviti L, Seow Y, Yin H, Betts C, Lakhal S, Wood MJ. Delivery of siRNA to the mouse brain by systemic injection of targeted exosomes. *Nat Biotechnol.* 2011;29(4):341-5.

[171] Momen-Heravi F, Bala S, Bukong T, Szabo G. Exosome-mediated delivery of functionally active miRNA-155 inhibitor to macrophages. *Nanomedicine.* 2014;10(7):1517-27.

[172] de Rivero Vaccari JP, Brand F, 3rd, Adamczak S, Lee SW, Perez-Barcena J, Wang MY, et al., Exosome-mediated inflammasome signaling after central nervous system injury. *J Neurochem.* 2016;136 Suppl 1:39-48.

[173] Lener T, Gimona M, Aigner L, Borger V, Buzas E, Camussi G, et al., Applying extracellular vesicles based therapeutics in clinical trials - an ISEV position paper. *J Extracell Vesicles.* 2015;4:30087.

[174] Choi SH, Kim YH, Hebisch M, Sliwinski C, Lee S, D'Avanzo C, et al., A three-dimensional human neural cell culture model of Alzheimer's disease. *Nature.* 2014;515(7526):274-8.

[175] Carlson AL, Bennett NK, Francis NL, Halikere A, Clarke S, Moore JC, et al., Generation and transplantation of reprogrammed human neurons in the brain using 3D microtopographic scaffolds. *Nat Commun.* 2016;7:10862.

[176] Rosen DR. Mutations in Cu/Zn superoxide dismutase gene are associated with familial amyotrophic lateral sclerosis. *Nature.* 1993;364(6435):362.

[177] Forsberg K, Jonsson PA, Andersen PM, Bergemalm D, Graffmo KS, Hultdin M, et al., Novel antibodies reveal inclusions containing non-native SOD1 in sporadic ALS patients. *PLoS One.* 2010;5(7):e11552.

[178] Neumann M, Sampathu DM, Kwong LK, Truax AC, Micsenyi MC, Chou TT, et al., Ubiquitinated TDP-43 in frontotemporal lobar degeneration and amyotrophic lateral sclerosis. *Science.* 2006;314 (5796):130-3.

[179] Sreedharan J, Blair IP, Tripathi VB, Hu X, Vance C, Rogelj B, et al., TDP-43 mutations in familial and sporadic amyotrophic lateral sclerosis. *Science.* 2008;319(5870):1668-72.

[180] Mackenzie IR, Bigio EH, Ince PG, Geser F, Neumann M, Cairns NJ, et al., Pathological TDP-43 distinguishes sporadic amyotrophic lateral

sclerosis from amyotrophic lateral sclerosis with SOD1 mutations. *Ann Neurol.* 2007;61(5):427-34.

[181] Vance C, Rogelj B, Hortobagyi T, De Vos KJ, Nishimura AL, Sreedharan J, et al., Mutations in FUS, an RNA processing protein, cause familial amyotrophic lateral sclerosis type 6. *Science.* 2009;323(5918):1208-11.

[182] Zou ZY, Cui LY, Sun Q, Li XG, Liu MS, Xu Y, et al., De novo FUS gene mutations are associated with juvenile-onset sporadic amyotrophic lateral sclerosis in China. *Neurobiol Aging.* 2013;34(4):1312 e1-8.

[183] Greenway MJ, Andersen PM, Russ C, Ennis S, Cashman S, Donaghy C, et al., ANG mutations segregate with familial and 'sporadic' amyotrophic lateral sclerosis. *Nat Genet.* 2006;38(4):411-3.

[184] Maruyama H, Morino H, Ito H, Izumi Y, Kato H, Watanabe Y, et al., Mutations of optineurin in amyotrophic lateral sclerosis. *Nature.* 2010;465(7295):223-6.

[185] van Blitterswijk M, van Vught PW, van Es MA, Schelhaas HJ, van der Kooi AJ, de Visser M, et al., Novel optineurin mutations in sporadic amyotrophic lateral sclerosis patients. *Neurobiol Aging.* 2011;33(5):1016 e1-7.

[186] Renton AE, Majounie E, Waite A, Simon-Sanchez J, Rollinson S, Gibbs JR, et al., A hexanucleotide repeat expansion in C9ORF72 is the cause of chromosome 9p21-linked ALS-FTD. *Neuron.* 2011;72(2):257-68.

[187] Lattante S, Conte A, Zollino M, Luigetti M, Del Grande A, Marangi G, et al., Contribution of major amyotrophic lateral sclerosis genes to the etiology of sporadic disease. *Neurology.* 2012;79(1):66-72.

In: New Developments in Astrocytes Research ISBN: 978-1-63485-561-7
Editor: Leyla Estrada

Chapter 4

ASTROCYTE'S RESPONSE TO STRESS AND INJURY USING FGF-1/APOE/HDL SYSTEM

Jin-ichi Ito** and Makoto Michikawa
Department of Biochemistry,
Nagoya City University Graduate School of Medical Sciences
Mizuho-cho, Mizuho-ku, Nagoya, Japan

ABSTRACT

Astrocytes are most abundant glial cells in the brain and their functions are widely known. Astrocytes support morphogenesis and localization of neurons in the brain through the formation of astroglial networks. Selective molecular transport between inside and outside of the central nervous system (CNS) mediated by the blood-brain barrier (BBB) is regulated by astrocytes to control the function of endothelial cells that construct the brain capillary. Astrocytes digest and remove excess neurotransmitters released from the pre-synaptic neurons for the control of neurotransmitter's concentration in the synaptic cleft. Furthermore, astrocytes produce and secrete neurotransmitter-like molecules such as glutamine and D-serine as glio-transmitters. Recently, there are a number of reports that astrocytes contribute to neuronal protection by producing antioxidants such as glutathione and metallothionein, apolipoprotein E-containing high-density lipoprotein-like particles (apoE/HDL).

* Telephone: +81-52-853-8140, FAX, +81-52-841-3480, E-mail: jitoh@med.nagoya-cu.ac.jp.

We have found that the release of fibroblast growth factor 1 (FGF-1) is enhanced from rat astrocytes exposed to the gentle oxidative stress without inducing apoptosis. The gentle oxidative stress suppresses transiently syntheses of lipids such as cholesterol, phosphatidylcholine, and sphingomyelin, and enhances lipid release from the cell surface, resulted in increasing permeability of the plasma membrane to induce release of FGF-1 in addition to the cytosolic proteins such as HSP110, HSP90, and HSP70. FGF-1 enhances the generation of apoE/HDL through the activation of at least three kinds of signaling in stress-less and healthy rat astrocytes without oxidative stress. The apoE/HDL works to protect astrocytes and neurons from oxidative stress. Astrocytes are much more sensitive to oxidative stress for the release of FGF-1 and cytosolic proteins than other cell strains. Astrocytes may have a higher performance of sensor against oxidative stress to enhance FGF-1 release than other cell strains. These findings suggest that astrocytes have a potential system using an FGF-1/apoE/HDL to protect neural cells from oxidative stress in the brain.

Keywords: astrocytes, FGF1, apoE, HDL, oxidative stress, cholesterol

ABBREVIATIONS

FGF-1	fibroblast growth factor 1
apoE	apolipoprotein E
HDL	high-density lipoprotein
LDL	low-density lipoprotein
BBB	blood-brain barrier
CNS	central nervous system
CSF	cerebrospinal fluid
ROS	reactive oxygen species
SOD	superoxide dismutase
LRP	LDL receptor-related protein
NGF	nerve growth factor
BDNF	brain-derived neurotrophic factor

1. INTRODUCTION

Oxidative stress constitutes a great menace to the function and viability of brain. Oxygen is necessary for all animals to be alive. All animals must produce continuously adenosine triphosphate (ATP) in mitochondria as vivo energy for keeping the metabolism to consume oxygen and glucose. Especially, the brain consumes severely oxygen to keep functional activities of the CNS. The brain is thought to be much higher dependent on the oxidation of glucose than other organs. Consumption of oxygen facilitates to produce reactive oxygen species (ROS) through various chemical reactions between oxygen molecule and organic compounds in the brain. ROS is a general term of oxygen compounds with highly reactive activities. Accordingly, the brain produces many kinds of oxidized lipids with facility because the brain is occupied by more than 50% of lipids including abundant amounts of unsaturated fatty acid. Oxygen is a two-edged sword for oxygen-consuming animals. There are surely many reports suggesting that oxidative stress gives rise to various brain diseases, or that lots of brain diseases are concomitant with oxidative stress. Several diseases of brain such as Parkinsonism and Alzheimer's disease are seemingly accompanied by oxidative stress (Hernandes and Britto, 2012).

The brain consumes highly oxygen and produces various ROS at high level, resulted that neural cells undergo easily oxidative stress. ROS produces not only many kinds of peroxide lipids in the lipid-rich brain but also peroxides of amino acids and bases. It is, however, known that there are several kinds of protection systems against oxidative stress in the brain. Anti-oxidants such as ascorbic acid and α-tocophenol derived from foods protect the brain against oxidative stress induced by superoxide in addition to antioxidant enzymes such as superoxide dismutase (SOD) and glutathione peroxidase. Astrocytes work actively to protect the brain against oxidative stress also. Astrocytes produce and release glutathione (GSH) and metallothionein to protect neurons against oxidative stress (Dringen et al., 1999; Wang and Cynader, 2000). Furthermore, astrocytes secrete ATP to activate a specific ATP receptor, P2Y1, in the autocrine manner, and then enhance the expression and release of thioredoxin reductase (TrxR) in/from astrocytes for protection of the brain against oxidative stress (Eftekharpour et al., 2000).

It has been gradually known that astrocytes participate in protection and repair of the brain from injuries and stresses. Ignatius et al., reported that the synthesis of 37-kDa glycoprotein is dramatically increased after injury of the rat sciatic and optic nerves (Ignatius et al., 1986). They demonstrated that the 37 kDa protein induced in injured nerves is identical to apoE. Oligodendroglias, furthermore, express apoE after injury of rat optic nerve also (Stoll et al., 1989). Thus, it is sure that the production of apoE is induced after the brain nerve injury. It is believed that apoE is produced mainly by astrocytes and partly by microglias and distributed in the brain as a most abundant CSF lipoprotein, apoE/HDL (Chiba et al., 1991; Ito et al., 1999). Because neural cells in the brain are isolated by the blood-brain barrier (BBB) from large molecules and macromolecular complexes such as lipoproteins in the plasma, it is thought that the neural cells hardly incorporate cholesterol through the interaction with plasma lipoproteins. Therefore, apoE/HDL is important for intercellular cholesterol transport in the brain and astrocytes are important glial cells as apoE/HDL-producing cells.

FGF-1 is known generally released under the stressful condition. We found that rat astrocytes release FGF-1 under gentle oxidative stress condition induced by H_2O_2, and furthermore that the production and release of FGF-1 are promoted in/from rat astrocytes during the long-term culture accompanied with oxidative stress (Ito et al., 2013; Nagayasu et al., 2014). FGF-1 released from astrocytes under oxidative stress enhances the generation of apoE/HDL in stress-less astrocytes likely in the manner of paracrine action. Hayashi et al., showed that the apoE/HDL derived from rat astrocytes protects neurons from apoptosis induced by withdrawal of trophic additives (Hayashi et al., 2007). Thus, the combination action using FGF-1, apoE, and HDL (FGF-1/apoE/HDL system) derived from astrocytes has significant roles for safety of the brain to protect neural cells from various stresses.

2. FGF-1 Release from Astrocytes under Oxidative Stress

Ogura et al., reported that subtoxic levels of oxidative stress induced by 100 – 250 μM hydrogen peroxide (H_2O_2) promote the expressions of neurotrophic factors, brain-derived neurotrophic factor (BDNF) and progranulin in murine hippocampal cells to protect neuronal cells (Ogura et al., 2014). We also found that 100 μM H_2O_2 enhances only FGF-1 release from rat astrocytes but not FGF-1 expression without inducing apoptosis (Ito et al., 2013). On the other hand, rat astrocytes enhance both production and release of FGF-1 during long-term culture for one month (Nagayasu et al., 2008). Rat astrocytes are conventionally prepared by culture of 18-day fetal rat brain cerebrum for one week in 10% fetal calf serum-containing F-10 medium after removal of the meninges and cutting into small pieces as a primary culture, and then more one-week culture as a secondary culture after trypsinization of the primary culture. This conventional culture of rat astrocytes (W/W cells) hardly induces FGF-1 production during the one-week secondary culture. The prolongation of secondary culture to one month from one-week, however, enhances production and release of FGF-1 and apoE in rat astrocytes (W/M cells). The W/M cells under long-term culture for one-month indicate much higher levels of Bax in the mitochondria as a maker protein for mitochondrial stress and superoxide dismutase 1 (SOD1) in the cytosol as an oxidative stress marker than conventional rat astrocytes (W/W cells). W/M cells have no susceptibility to FGF-1 activity to phosphorylate protein kinases such as Akt, MEK, and ERK similarly to W/W cells treated with H_2O_2 (Nagayasu et al., 2014). These findings suggest that the long-term culture for one month loads mitochondrial stress and oxidative stress on rat astrocytes.

As FGF-1 has no amino-terminal signal sequence as well as FGF-2, FGF-1 is thought synthesized and localized in the cytosol of FGF-1-producing cells. Accordingly, the FGF-1 release to the extracellular space is unlikely carried out through the classical intracellular transport system via endoplasmic reticulum (ER)/Golgi pathway after the biosynthesis. Endogenous FGF-1 appears to be localized as a high molecular weight complex composed of FGF-1, p40 extravascular domain of synaptotagmin (Syn)-1, and S100A13 in neural tissues and seems to be released to the extracellular space as the complex under the stressful condition such as oxidative stress, heat shock, hypoxia, and serum starvation (Jackson et al., 1992; Mouta et al., 2001; Shin et al., 1996). We showed also that rat astrocytes release FGF-1 under oxidative stress and mitochondrial stress during long-term culture. However, the mechanism underlying FGF-1 release is not fully understood at present despite of many previous investigations.

Strangely, the W/M cells under oxidative stress enhances not only release of FGF-1 but also expression of FGF-1 mRNA unlike W/W cells treated with H_2O_2. As described above, oxidative stress enhances only FGF-1 release from rat astrocytes without increasing FGF-1 production. The syntheses of lipids such as cholesterol, sphingomyelin, and phosphatidylcholine in W/W cells are suppressed by gentle oxidative stress induced by the treatment with 100 μM H_2O_2 for 10 min. The gentle oxidative stress enhances release of lipids from the cell surface of rat astrocytes also, accompanied with the release of cytosolic proteins such as HSP110, HSP90, HSP70, and β-actin. Antioxidants such as ascorbic acid and ebselen suppress the release of cytosolic proteins induced by H_2O_2 treatment. Furthermore, exogenously added molecule such as trypsin or Rhodamin-phalloidin is permitted with facility to permeate into the intracellular side of rat astrocytes treated with gentle oxidative stress, resulted in Rhodamin-staining of intracellular actin filaments and proteolysis of cellular proteins by trypsin. The pretreatment of rat astrocytes with lipoproteins such as LDL and HDL reduced H_2O_2–induced release of not only cytosolic proteins but also FGF-1. These findings suggest that the gentle oxidative stress with H_2O_2 induces impairment in lipid structure and higher permeability of the plasma membrane to rat astrocytes. The change of lipid metabolism may participate in the increase of membrane permeability induced by H_2O_2.

The sensitivities to H_2O_2 of other cell types such as mouse Balb3T3, bovine endothelial cells, and mouse hepatocytoma HepG2 cells are very low for release of cytosolic proteins in comparison with rat astrocytes (Ito et al., 2013). Although treatment with only 20 μM H_2O_2 enhances the release of cytosolic proteins from rat astrocytes, the release from mouse Balb3T3 cells is slightly enhanced at 300 μM H_2O_2. Both bovine endothelial cells and HepG2 cells failed to enhance the release even at 500 μM H_2O_2. This finding suggests that astrocytes are more sensitive to oxidative stress using H_2O_2 for the release of cytosolic proteins containing FGF-1. Astrocytes may have a sensor system with higher performance to oxidative stress than those of other cell lines for the release of FGF-1 to protect the brain against oxidative stress. Furthermore, it is thought that the FGF-1 released from stressful astrocytes functions as a stimulant of apoE/HDL generation in stress-less astrocytes in the manner of paracrine action. Oxidative stress may be one of candidates which triggers FGF-1 release from astrocytes in the brain.

3. Anti-Oxidant Effect of FGF-1

Does FGF-1 have an anti-oxidant function in the brain? The FGF-1 production is found in all motor neurons and primary sensory neurons in the mesencephalon and in a small number of glial cells in the white matter of control and injuried brain (Eckenstein et al., 1994). FGF-1 is localized highly in motor neurons in the spinal cord and released by injury in the spinal cord also (Cassina et al., 2005). The released FGF-1 activates astrocytes to enhance the expression and secretion of nerve growth factor (NGF) through FGF receptor-1. NGF has a neuroprotective function to support differentiation and survival of neurons, although it functions to remove damaged neurons and glia also. Indeed, Cassina et al., found that FGF-1 secreted by motor neurons in amyotrophic lateral sclerosis (ALC) activates and induces astrocytes to reactive strocytes to produce NGF, which induces apoptosis of motor neurons through the interaction between NGF and p75 neurotrophin receptor ($p75^{NTR}$). Interestingly, mutant superoxide dismutase 1 (SOD1)-transgenic mice exhibit more greatly the expression of FGF-1 in motor neurons due to oxidative stress (Cassina et al., 2005). This finding suggests that SOD1 plays protection role of neurons from ALS-related oxidative stress. Furthermore, increasing FGF-1 expression is observed in reactive astrocytes in the gray and white matters of human brain of Alzheimer's disease and Huntington's disease (Tooyama et al., 1993; Tooyama et al., 1991). Stresses have seemingly significant influence on FGF-1 production in the brain, suggested that FGF-1 participates greatly in brain protection from stresses and injuries.

FGF-1 induces the expression of heme oxygenase-1 (HO-1) in cultured rat spinal cord astrocytes through the induction of mRNA and protein production of nuclear factor erythroid 2-related factor 2 (Nrf2) (Vargas et al., 2005). Heme oxygenase is a microsomal enzyme to catalyze the degradation of heme groups to yield biliverdin, iron, and carbon monoxide, followed by conversion of biliverdin to antioxidant bilirubin by biliverdin reductase. The induction of HO-1 is thought to contribute in protection from oxidative damage. There are reports showing that the HO-1 expression is increased in neurodegenerative diseases involving glial activation such as Alzheimer's disease and Parkinson disease (Schipper, 2011). The mechanism underlying FGF-1 action to enhance production of NGF and HO-1 is not well known with relevance to cell signaling pathways at present. Thus FGF-1 potently activates astrocytes and exerts indirect neuroprotection after stress or injuty.

We observed that the treatment with amyloid β-42 but not amyloid β-40 enhances the release of not only FGF-1 from rat astrocytes but also cytosolic proteins such as HSP70, HSP90, and HSP110 from rat neurons. Furthermore, we found that the pretreatment of rat astrocytes with FGF-1 suppresses the release of cytosolic proteins from the cells treated with H_2O_2. Rat astrocytes restore resistance to oxidative stress induced by H_2O_2 after the pretreatment with lipoproteins such as LDL and HDL dependently on the increase in lipid content of plasma membrane. As FGF-1 increases syntheses of lipids such as cholesterol, sphingomyelin, and phosphatidylcholine in rat astrocytes, rat astrocytes treated with FGF-1 may become resistant to oxidative stress (Ito et al., 2015). However, It is unclear whether FGF-1 suppresses directly the protein release from rat astrocytes or suppresses it indirectly through the increasing apoE/HDL generation. It is an interesting subject whether the apoE/HDL generated by FGF-1-treated astrocytes participates in the FGF-1-derived actions to protect neural cells from stresses and injuries.

4. APOE/HDL GENERATION OF ASTROCYTES

It has been reported that the apoE production is increased in the sites of injury and stress in the brain (Snipes et al., 1986), although it is unclear whether the apoE accumulation in the wound site is significant for protection of neuronal cells from injury and stress or for removal and transport of lipids from apoptotic cells. Hayashi et al., reported that a lipidated apoE protects neuronal cells from apoptosis induced by withdrawal of trophic additives (Hayashi et al., 2007; Hayashi et al., 2009), suggesting that apoE/HDL has an important role to protect of function and survival of neuronal cells in the brain. How is an apoE/HDL generation regulated in the brain?

In the brain, apoE is synthesized by mainly astrocytes and partially microglias (Chiba et al., 1991; Ito et al., 1999). Rat astrocytes generate not only cholesterol-rich apoE/HDL (1.08 – 1.12 g/ml) as major brain lipoproteins but also phosphatidylcholine-rich apoA-I/HDL (1.12 – 1.17 g/ml) through the interaction with exogenous apoA-I secreted by brain capillary endothelial cells. This finding suggests that astrocytes-derived apoE/HDL and apoA-I/HDL have different roles for brain cholesterol homeostasis dependently on different properties of cholesterol content in each lipoprotein. As it is known that apoA-I stimulates apoE secretion in foam cell macrophages (Rees et al., 1999), exogenous apoA-I or apoA-I/HDL influences positively on apoE secretion from astrocytes or apoE/HDL generation in the brain. It is unclear at present whether apoE is released as a lipid-free apolipoprotein or as a lipoprotein along with lipids from astrocytes. We have studied mechanism of apoE secretion using astrocytes prepared from fetal rat brain. The cholesterol released by rat astrocytes is generally recovered along with the secreted apoE with densities of 1.08 – 1.12 g/ml in the conditioned medium as apoE/HDL. The apoE secretion from rat astrocytes is initiated within 30 min after start of the biosynthesis, along with the cholesterol prebiosynthesized several days ago but not the newly synthesized cholesterol. Strangely, the newly synthesized ^{14}C-labeled cholesterol is recovered in the HDL fraction with the higher density fraction (1.12 – 1.17 g/ml) than the density of apoE/HDL fraction (1.08 – 1.12 g/ml) in the conditioned medium after the 24 h-incubation of rat astrocytes with ^{14}C-acetate for the biosynthesis of ^{14}C-labeled cholesterol (Ito et al., 2014). These findings suggest that the endogenous apoE carries out the previously synthesized cholesterol but not newly synthesized one to the extracellular space unlike exogenously added lipid-free apoE. This implies that the newly synthesized cholesterol is released as lipid-poor HDL through the association with some kind of apolipoprotein except for apoE in the conditioned medium. It is known that apoJ secreted by astrocytes is recovered with higher density fraction as a lipid-poor HDL in the conditioned medium (Fagan et al., 1999). As exogenously added lipid-free apoE and apoA-I are able to enhance the release of both previously synthesized cholesterol and newly synthesized one to generate HDL with lower density of 1.11 – 1.12 g/ml, the newly biosynthesized apoE may be secreted as a lipidated apoE but not a lipid-free apoE. The mechanism for apoE/HDL generation mediated by endogenous apoE appears to be different from that for apoE/HDL generation mediated by exogenously added lipid-free apoE. It is very interesting whether apoE is intracellularly lipidated in astrocytes.

There are many reports to suggest that ABCA1 participates in the apoE-mediated HDL generation. Wahrle et al., reported that the apoE levels are decreased in both the cortex (80% reduction) and the CSF (98% reduction) of ABCA1-KO mouse (Wahrle et al., 2004). The CSF in ABCA1-KO mouse significantly reduces cholesterol level as well as small sizes of apoE-containing lipoproteins. We demonstrated the intracellular binding of endogenous apoE and ABCA1 by immunoprecipitation (Ito et al., 2014). Even though endogenous apoE associates with intracellular ABCA1, the intracellular binding of these molecules may be very slight, because large parts of ABCA1 are localized in the plasma membrane. The apoE/HDL generated by ABCA1-KO mouse astrocytes is recovered with the higher density fraction in the conditioned medium than that generated by wild type of mouse astrocytes. The ABCA1 is distributed in both fractions of the caveolin-1-rich fraction containing lipid raft domain and non-caveolin-1-fractoin with higher density than the caveolin-1-rich fraction. While exogenously added apoA-I is associated with ABCA1 in the non-caveolin-1 fraction, endogenous apoE is associated with ABCA1 in the caveolin-1-rich fraction (Ito et al., 2011; Ito et al., 2014). Interestingly, the endogenous apoE in ABCA1-KO astrocytes is localized in non-caveolin-1 fraction. The cholesterol content of HDL generated by astrocytes may be dependent on the distribution site of ABCA1 that associates with apolipoprotein in the plasma membrane.

5. FGF-1-INDUCED ENHANCEMENT OF APOE/HDL GENERATION

We have ever observed that astrocytes treated with FGF-1 enhance apoE/HDL generation (Ueno et al., 2002). Long-term cultured rat astrocytes

(W/M cells) enhance to mRNA expression and protein synthesis of FGF-1, although the W/M cells have a lower sensitivity to FGF-1 (Nagayasu et al., 2008). The W/M cells under oxidative stress induced by H_2O_2 enhances release FGF-1, which up-regulates both syntheses and release of cholesterol and pholipids in stress-less rat astrocytes (W/W cells) through the increasing production of apoE but not stressful W/M cells. After injuring with liquid nitrogen in mouse brain, immunohistchemical analysis of the lesions revealed that the FGF-1 production is increased by the day 2, and then the apoE production is enhanced by the day 4 in astrocytes after the cryo-injury in the wild-type mouse brain. This finding suggests that the cryo-injury of mouse brain enhances FGF-1 production of astrocytes, followed by increasing apoE production seemingly to depend on FGF-1 production. Thus, we took notice of FGF-1-induced generation of apoE/HDL in astrocytes in order to understand the mechanism underlying protection of brain from stress and injury.

We demonstrated that purified FGF-1 enhances both biosyntheses and secretion of apoE and cholesterol to generate apoE/HDL as well as an FGF-1-like factor synthesized and released by rat astrocytes into the conditioned medium under oxidative stress induced by H_2O_2. FGF-1 activates both mitogen-activated protein kinase (MAP kinase) cascade through phosphorylation of MEK and ERK for enhancing cholesterol synthesis and phosphatidylinositide 3-OH kinase (PI-3 kinase)/Akt pathway through phosphorylation of Akt for intracellular apoE transport to the plasma membrane (Ito et al., 2005; Ito et al., 2007). An MEK inhibitor, U0126, suppresses FGF-1-induced cholesterol synthesis without inhibition of cholesterol release in rat astrocytes. The enhancement of cholesterol release by FGF-1 is neutralized by a PI3-kinase inhibitor, LY294002, but not by U0126. FGF-1 regulates positively cholesterol biosynthesis without enhancing cholesterol release in apoE-KO mouse astrocytes, indicating a signaling pathway independent of apoE biosynthesis upregulation. This finding, furthermore, supports that FGF-1 enhances cholesterol release from astrocytes dependently on apoE release.

FGF-1 enhances Src phosphorylation at Tyr416 in the cytosol, and both the FGF-1-induced signal transduction and generation of apoE/HDL are regulated by non-receptor-type tyrosin protein kinase such as Src (Nishida et al., 2009). Src inhibitors such as SU6656 and PP1 suppress FGF-1-induced cell signaling such as phosphorylation of Akt, MEK, and ERK. The down-regulation of Src gene by Src SiRNA also reduces the FGF-1-induced cell signaling. These inhibitors, furthermore, suppress not only cholesterol synthesis but also apoE secretion in rat astrocytes treated with FGF-1. These findings suggest that FGF-1 activates signal pathway of MEK/ERK and PI-3 kinase/Akt through the activation of Src tyrosine kinase in rat astrocytes, resulted in enhancement of cholesterol synthesis and apoE secretion.

We found that the mRNA expression of cholesterol 25-hydroxylase is enhanced by FGF-1 and suppressed by inhibition of cholesterol syntheses with compactin in rat astrocytes (Lu et al., 2009). FGF-1 indeed enhanced the production 25-hydroxycholesterol, which up-regulates apoE gene expression as a LXRα activator oxysterol. The increase of LXRα mRNA was suppressed by inhibition of activation of FGF-1 receptor-1 or MEK/ERK cascade, which participates in the increase in cholesterol synthesis induced by FGF-1. The inhibition of PI3-kinase/Akt pathway had no effect on FGF-1-induced LXRα transcription. The increases of apoE mRNA transcription and apoE/HDL generation were both inhibited by down-regulation or inhibition of LXRα. Liang et al., showed that the LXR/RXR heterodimer mediates apoE mRNA expression and secretion in macrophages for cholesterol homeostasis (Liang et al., 2004). We have no idea whether FGF-1 enhances the transcription of retinoid X receptor (RXR) mRNA in astrocytes.

6. Anti-Stress Activities of apoE/HDL

It has been believed that the BBB has significant roles to protect function and viability of neural cells in the brain through the suppression of invasion of bacteria, virus, or drugs from the blood. The apoE/HDL generated by astrocytes appears to up-regulate the structure and function of BBB constructed by brain capillary endothelial cells, as apoE-deficiency induces BBB disfunction. Methia et al., reported that the spontaneous leakage of injected Evans blue dye in the brain of apoE-deficient mice is 70% higher than that in the wild type brain after injury induced by dry ice (Methia et al., 2001). Nishituji et al., found that apoE3 contributes more greatly to BBB function than apoE4 does (Nishitsuji et al., 2011). The apoE3/HDL enhances occuludin phosphorylation to enhance tight junction integrity through the interaction with LDL receptor-related protein1 (LRP1), suggesting that the tight junction of brain capillary endothelial cells is regulated by apoE in an isoform-dependent manner. Furthermore, apoE deficiency promotes the BBB disruption through up-regulation of the mRNA expression of matrix metalloproteinase (MMP)-9, which is known to degrade major components of basement membrane of the BBB such as type IV collagen, laminin, and fibronectin (Zheng et al., 2014). Thus apoE/HDL in the brain contributes to preserve the function of BBB.

Astrocytes produce and secrete transforming growth factor-β (TGF-β) to activate endothelial cells to increase tight junction for strength of vascular structures (Tao-Cheng et al., 1987; Tran et al., 1999). The TGF-β-induced activation of MAP kinase cascade and SMAD signaling pathway up-regulates tight junction protein claudin-1. Furthermore, it is known that TGF-β enhances the expression of apoE gene through the activation of c-Jun N-terminal kinase, p38 kinase, and casein kinase 2 (Singh and Ramji, 2006). Thus, the activity of TGF-β to strengthen tight junction of endothelial cells may be dependent on enhancement of apoE production and apoE/HDL generation also.

Hayashi et al., reported that neuronal apoptosis through the activation of calcineurin induced by withdrawal of tropic additives is blocked by the apoE/HDL generated in astrocytes (Hayashi et al., 2009). They demonstrated that the apoE/HDL binds to LRP1 in the neurons, followed by activation of phospholipase Cγ1, an increase in diacylglyceride production to activate protein kinase Cδ, and finally suppression of GSK3β phosphorylation. The apoE/HDL-induced series of reactions suppress the activation of calcineurin. This protection activity of apoE/HDL cannot be replaced by only lipid, apoA-I/HDL, apoJ/HDL, or lipid-free apoE. The incorporation of apoE/HDL by neurons is not required for neuronal protection from this type of apoptosis, suggesting that suppression of neuronal apoptosis is mediated through the specific signal transduction caused by the lipidated apoE but not lipid-free apoE.

We found that the surface size of brain wound caused by liquid nitrogen is shrunk more rapidly in the wild-type mouse brain than in the apoE-knock out (apoE-KO) mouse brain, although the FGF-1 production is also observed by the day 2 in apoE-knockout mice brain. It is interesting that the wound healing delayed in the apoE-KO mouse brain despite of same level of FGF-1 production in apoE-KO mouse brain in comparison with the wild-type mouse brain. Immunohistochemical analysis of the lesion sites revealed that the FGF-1 production is increased in astrocytes by the day 2 after the injury in both wild-type and apoE-KO mice brains, while the production of apoE is enhanced by the day 4 in astrocytes of the wild-type mouse. Our experimental results indicate that the recovery from cryo-injury in the brain appears to depend on FGF-1-induced apoE/HDL generation rather than FGF-1. This finding suggests that the apoE/HDL generated by astrocytes contributes to recover the brain from the injury. The enhancement of apoE production for recovery of brain from injury may be one of important roles of FGF-1. We already demonstrated that FGF-1 has an activity to enhance the secretion of apoE and the generation of apoE/HDL in cultured rat astrocytes. It is suggested that the FGF-1 produced by astrocytes enhances apoE/HDL generation of astrocytes in the autocrine/paracrine manner in the injured brain.

There appears to have been no report showing that Alzheimer's disease enhances apoE production. However, a large-scale study of CSF samples from both normal individuals and Alzheimer's disease patients appears to indicate that the level of apoE in CSF is positively associated with the level of Aβ42 in CSF independently of apoE4 genotype, namely, the level of apoE in the CSF is higher in high CSF Aβ42 than in low CSF Aβ42 (Cruchaga et al., 2012). Alzheimer's disease is known to be accompanied with oxidative stress and to enhance the FGF-1 production in the brain. We found that oxidative stress enhances FGF-1 release from rat astrocytes (Ito et al., 2013). Accordingly, Alzheimer's disease may enhance apoE/HDL generation of astrocytes in the brain. It is known that some kinds of neurotrophins such as NGF and BDNF have activity to protect neuronal cells from oxidative stress. These factors activate MAP kinase and phosphatidylinositol 3-kinase (PI3K) cascades through Trk receptors with tyrosine kinase activity as their receptors (Wooten et al., 2000). It is interesting whether these neurotrophins enhance apoE/HDL generation of neural cells, as many evidences support that the apoE/HDL is important for protection of neural cells from stress and injury in the brain.

CONCLUSION

There are gradually increasing evidences to indicate apoE/HDL functions in not only intercellular cholesterol transport in the brain but also protection of the brain from various injuries and stresses. In the brain, apoE/HDL is mainly generated by astrocytes. The existence of astrocytes, therefore, is thought very important for cholesterol homeostasis and protection of the brain. Astrocytes under stressful condition produce and release FGF-1, which enhances apoE/HDL generation in astrocytes in the autcrine/paracrine manner, This means that astrocytes have a system to self-control apoE/HDL generation in response to various stressful external environments. Interestingly, astrocytes are more sensitive to oxidative stress for the release of cytosolic proteins containing FGF-1 than other cell strains. It will be more clear by further studies that FGF-1/apoE/HDL system derived from astrocytes is important for brain protection from stresses and injuries.

REFERENCES

Cassina, P., et al., 2005. Astrocyte activation by fibroblast growth factor-1 and motor neuron apoptosis: implications for amyotrophic lateral sclerosis. *J. Neurochem.* 93, 38-46.

Chiba, H., et al., 1991. Apolipoproteins in rat cerebrospinal fluid: a comparison with plasma lipoprotein metabolism and effect of aging. *Neurosci. Letters.* 133, 207-210.

Cruchaga, C., et al., 2012. Cerebrospinal fluid apoE levels: an endophenotype for genetic studies for Alzheimer's disease. *Human Mol. Gen.* 21, 4558-4571.

Dringen, R., Pfeiffer, B., Hamprecht, B., 1999. Synthesis of the antioxidant glutathione in neurons: supply by astrocytes of CysGly as precursor for neuronal glutathione. *J. Neurosci.* 19, 562-569.

Eckenstein, F. P., et al., 1994. Distribution of acidic and basic fibroblast growth factors in the mature, injured and developing rat nervous system. *Prog. Brain Res.* 103, 55-64.

Eftekharpour, E., Holmgren, A., Juurink, B.H., 2000. Thioredoxin reductase and glutathione synthesis is upregulated by t-butylhydroquinone in cortical astrocytes but not in cortical neurons. *Glia.* 31, 241-248.

Fagan, A. M., et al., 1999. Unique lipoproteins secreted by primary astrocytes from wild type, apoE (-/-), and human apoE transgenic mice. *J. Biol. Chem.* 274, 30001-30007.

Hayashi, H., et al., 2007. Apolipoprotein E-containing lipoproteins protect neurons from apoptosis via signaling pathway involving low-density lipoprotein receptor-related protein-1. *J. Neurosci.* 27, 1933-1941.

Hayashi, H., et al., 2009. Protection of neurons from apoprosis by apolipoprotein E-containing lipoproteins does not require lipoprotein uptake and involves activation of phospholipase Cgamma1 and inhibition of calxineurin. *J. Biol. Chem.* 284, 29605-29613.

Hernandes, M. S., Britto, L. R. G., 2012. NADPH oxidase and neurodegeneration. Current Neuropharmacol. 10, 321-327.

Ignatius, M.J., et al., 1986. Expression of apolipoprotein E during nerve degeneration and regeneration. *Proc. Natl. Acad. Sci. USA.* 83, 1125-1129.

Ito, J.-i., et al., 1999. Differential generation of high-density lipoprotein by endogenous and exogenous apolipoproteins in cultured fetal rat astrocytes. *J. Neurochem.* 72, 2362-2369.

Ito, J.-i., et al., 2005. Astrocytes produce and secrete FGF-1, which promotes the production of apoE-HDL in a manner of autocrine action. *J. Lipid Res.* 46, 679-686.

Ito, J.-i., et al., 2007. Mechanism for FGF-1 to regulate biogenesis of apoE-HDL in astrocytes. *J. Lipid Res.* 48, 2020-2027.

Ito, J.-i., et al., 2011. ApoA-I enhances generation of HDL-like lipoproteins through interaction between ABCA1 and phospholipase Cgamma in rat astrocytes. *Biochim. Biophys. Acta.* 1811, 1062-1069.

Ito, J.-i., et al., 2013. Enhancement of FGF-1 release along with cytosolic proteins from rat astrocytes by hydrogen peroxide. *Brain Res.* 1522, 12-21.

Ito, J.-i., et al., 2014. Astrocyte's endogenous apoE generates HDL-like lipoproteins using previously synthesized cholesterol through interaction with ABCA1. *Brain Res.* 1570, 1-12.

Ito, J.-i., et al., 2015. Biochemical properties in membrane of rat astrocytes under oxidative stress. *Brain Res.* in press.

Jackson, A., et al., 1992. Heat shock induces the release of fibroblast growth factor 1 from NIH3T3 cells. *Proc. Natl. Acad. Sci. USA*. 89, 10691-10695.

Liang, Y., et al., 2004. A liver X receptor and retinoid X receptor heterodimer mediates apolipoprotein E expression, secretion and cholesterol homeostasis in astrocytes. *J. Neurochem.* 88, 623-634.

Lu, R., et al., 2009. FGF-1 induces expression of LXR and production of 25-hydroxycholesterol to upregulate the apoE gene in rat astrocytes. *J. Lipid Res.* 50, 1156-1164.

Methia, N., et al., 2001. ApoE deficiency compromises the blood brain barrier. *Molec. Med.* 7, 810-815.

Mouta, C., et al., 2001. The comparative release of FGF1 by hypoxia and temparature stress. *Growth Factors.* 18, 277-285.

Nagayasu, Y., et al., 2008. Reactivity of astrocytes to fibroblast growth factor 1 for biogenesis of apolipoprotein E-high density lipoprotein is down-regulated by long-time secondary culture. *J. Biochem.* 143, 611-616.

Nagayasu, Y., et al., 2014. Increasing cellular level of phosphatidic acid enhances FGF-1 production in long term-cultured rat astrocytes. *Brain Res.* 1563, 31-40.

Nishida, T., et al., 2009. FGF-1-induced reactions for biogenesis of apoE-HDL are mediated by Src in rat astrocytes. *J. Biochem.* 146, 881-886.

Nishitsuji, K., et al., 2011. Apolipoprotein E regulates the integrity of tight junctions in an isoform-dependent manner in an in vitro blood-brain barrier model. *J. Biol. Chem.* 286, 17536-17542.

Ogura, Y., et al., 2014. Subtoxic levels of hudrogen peroxide induce brain-derived neurotrophic factor expression to protect PC12 cells. *BMC Res.* Notes. 7, 840-847.

Rees, D., et al., 1999. Apolipoprotein A-I stimulates secretion of apolipoprotein E by foam cell macrophages. *J. biol. Chem.* 274, 27925-27933.

Schipper, H. M., 2011. Heme oxygenese-1 in Alzheimer disease: a tribute to Moussa Youdim. *J. Neural Transm.* 118, 381-387.

Shin, J., et al., 1996. Serum-stavation induces the extracellular appearance of FGF-1. *Biochim. Biophys. Acta.* 1312, 27-38.

Singh, N. N., Ramji, D. P., 2006. Transforming growth factor-beta-induced expression of the apolipoprotein E gene requires c-Jun N-terminal kinase, p38 kinase and casein kinase 2. *Arterioscler Thromb Vasc Biol.* 26, 1323-1329.

Snipes, G. J., et al., 1986. Nerve injury stimulates the secretion of apolipoprotein E by nonneuronal cells. *Proc. Natl. Acad. Sci. USA.* 83, 1130-1134.

Stoll, G., et al., 1989. Oligodendrocytes but not astrocytes express apolipoprotein E after injury of rat optic nerve. *Glia.* 2, 170-176.

Tao-Cheng, J. H., et al., 1987. Tight junction of brain endothelium in vitro are enhanced by astroglia. *J. Neurosci.* 10, 3293-3299.

Tooyama, I., et al., 1993. Acidic and basic fibroblast growth factor-like immunoreactivity in the striatum and midbrain in Huntington's disease. *Brain Res.* 610, 1-7.

Tooyama, L., et al., 1991. Acidic fibroblast growth factor-like immunoreactivity in brain of Alzheimer patients. *Neurosci. Lett.* 121, 155-158.

Tran, N. D., et al., 1999. Transforming growth factor-beta mediates astrocyte-specific regulation of brain endothelial anticoagulant factors. *Stroke.* 30, 1671-1678.

Ueno, S., et al., 2002. An acidic fibroblast growth factor-like factor secreted into the brain cell culture medium upregulates apoE synthesis, HDL secretion and cholesterol metabolism in rat astrocytes. *Biochim. Biophys. Acta.* 1589, 261-272.

Vargas, M. R., et al., 2005. Fibroblast growth factor-1 induces heme oxygenase-1 via nuclear factor erythroid 2-related factor 2 (Nrf2) in spinal cord astrocytes. *J. Biol. Chem.* 280, 25571-25579.

Wahrle, S. E., et al., 2004. ABCA1 is required for normal central nervous system apoE levels and for lipidation of astrocyte-secreted apoE. *J. Biol. Chem.* 279, 40987-40993.

Wang, X. F., Cynader, M. S., 2000. Astrocytes provide cystein to neurons by releasing glutathione. *J. Neurochem.* 74, 1434-1442.

Wooten, M. W., et al., 2000. Mapping of atypical protein kinase C within the nerve growth factor signaling cascade: relationship to differentiation and survival of PC12 cells. *Mol. Cell. Biol.* 20, 4494-4504.

Zheng, M., et al., 2014. ApoE-deficient promotes blood-brain barrier disruption in experimental autoimmune encephalomyelitis via alteration of MMP-9. *J. Mol. Neurosci.* 54, 282-290.

INDEX

A

D

E

F

J

K

L

M

N

O

T

U

V

W

Y

Z